The Tokyo Toilet

The Tokyo Toilet

Shibuya
Tokyo
Japan

ザ・トウキョウ・トイレット

TOTO出版

03-3407-9815／中村ビル 中村
@Park
P
空
P

代々木八幡宮
車路

代々木八幡宮
車路
参拝者用駐輪場
代々木八幡宮
参拝者用駐輪場
代々木八幡宮
参拝者用駐輪場
代々木八幡宮

七号通り
KEIO
七号通り
KEIO
七号通り
中野新橋
中野駅

305
明治通り
Meiji-dori Ave.
渋谷区神宮前一丁目
1 Jingumae Shibuya City

CONTENTS

THE TOKYO TOILETS

CONTENTS

PROJECT DESIGN

CONCEPTION AND CONSTRUCTION

SPECIAL CONTRIBUTION

誰もが快適に利用できる公共トイレを
Public Toilets That Everyone Can Use Comfortably

衛生面や安全性に不安があるなど、街なかの公共トイレには、これまで、あまりいい印象がなかったかもしれません。障がいのある方、乳幼児連れの方、サポートを必要とされる方々が安心して使える設備が備わっていない場合もあります。
公共トイレとは本来、どんな時にも、誰にでも、不安なく、快適に使ってもらえるものではないか。この主旨のもと、クリエイターとともに、安全で清潔で、誰もが快適に利用できる公共トイレを考え、地域住民をはじめとする多くの方々の協力で実現したのが「THE TOKYO TOILET」です。
東京・渋谷区内の街なかや公園内の公共トイレ、17カ所の既存施設を刷新。最新の衛生設備機器を備え、どの施設にも、車椅子利用者やオストメイトに対応し、利用者の性別を問わないユニバーサル・トイレを設けることとしています。
トイレは社会を映す鏡とも言われています。それは、トイレがこの社会に生きる、すべての人に必要な場所だからではないでしょうか。「THE TOKYO TOILET」が、多様性を認め合い、それぞれの個性を尊重しながらともに生きる、インクルーシブな社会づくりの一助になることを願っています。

Public toilets in cities may not have left a good impression in the past due to sanitary and safety concerns. Some are not able to accommodate people with disabilities, people accompanying infants, and people needing support.
Public toilets should be accessible, comfortable, and anxiety-free for everyone at all times. With this concept in mind, the project team, together with the creators, considered what sort of public toilet would be safe, clean, and comfortable for everyone to use, and "THE TOKYO TOILET" was realized with the cooperation of many people, including the community residents.
Seventeen existing public toilet facilities in the streets and parks of Shibuya Ward, Tokyo, have been renovated. All facilities are fitted with the latest sanitation equipment and have gender-free universal toilets accommodating wheelchair users and ostomates.
Toilets are said to be a mirror of a society. This is because toilets are a necessary place for everyone in any society. THE TOKYO TOILET facilities will hopefully help to create an inclusive society where people can live together with respect for diversity and individuality.

P.2：広尾東公園トイレ
P.4-5：東三丁目公衆トイレ
P.6-7：代々木八幡公衆トイレ
P.8-9：七号通り公園トイレ
P.10-11：神宮前公衆トイレ
P.14：恵比寿東公園トイレ

p.2: Hiroo Higashi Park Public Toilet
pp.4-5: Higashi Sanchome Public Toilet
pp.6-7: Yoyogi-Hachiman Public Toilet
pp.8-9: Nanago Dori Park Public Toilet
pp.10-11: Jingumae Public Toilet
p.14: Ebisu East Park Public Toilet

THE TOKYO TOILETS

代々木深町小公園トイレ

YOYOGI FUKAMACHI MINI PARK PUBLIC TOILET

デザイン：坂 茂　渋谷区富ヶ谷1丁目54番1号

Design : Shigeru Ban　1-54-1 Tomigaya Shibuya-ku

2020年7月完成。鉄骨造。左からユニバーサル、女性用、男性用トイレが並ぶ。外壁は「瞬間調光ガラス」で、利用者が中に入って鍵をかけると、透明から不透明に変化する。

Completed in July 2020. Steel construction. Universal, women's, and men's toilet rooms are lined up from left to right. The exterior walls are made of smart glass, which changes from transparent to opaque when a user enters and locks the door.

左：建築面積は約14m²。公園には砂場や遊具もあり、子ども連れの利用者も多い。右上：井の頭通りの歩道から見た外観、北側。右中：女性用トイレ内から公園を見る。右下：庇とサッシはステンレスの鏡面仕上げ。

Left: The building area is about 14 m². The park has a sandpit and playground equipment and attracts many visitors with children. Top right: Exterior view from the sidewalk on Inokashira Dori on the north side. Middle right: View of the park from inside the women's toilet room. Bottom right: The eaves and window frames made of mirror-finished stainless steel.

はるのおがわ
コミュニティパークトイレ

HARU-NO-OGAWA
COMMUNITY PARK PUBLIC TOILET

デザイン：坂 茂　渋谷区代々木5丁目68番1号

Design : Shigeru Ban　5-68-1 Yoyogi Shibuya-ku

左：ユニバーサル・トイレ側面。個室間の壁は鏡面仕上げで、未使用時はガラスを通して周辺の木々を映す。右上：ガラスが透明な状態の、未使用時の外観。施設背後まで見通せる。右下：不透明に変化した使用時の外観。

Left: Side view of the universal toilet room. The walls between the rooms are mirror-finished, reflecting the surrounding trees through the glass when not in use. Top right: The exterior view of the facility when not in use, with transparent glass, making the rear of the facility visible. Bottom right: The exterior view of the facility when in use, with the glass turned opaque.

流す大
流す小
TOTO
止
おしり
ビデ
乾燥
ノズルきれい
TOTO

女性用トイレ内。自動水栓の手洗い場も備えた個室完結型で、通常の公共トイレの個室よりもゆとりのある広さを確保している。

Interior view of the women's toilet room. It is a private room equipped with an automatic faucet wash basin and is more spacious than a regular public toilet cubicle.

左上：敷地北側から未使用時の男性用トイレを見る。個室内に小便器も備える。左下：ガラスが不透明の状態の、使用時の外観夕景。右：夜は施設全体が行灯のように公園を照らす。

Top left: View of the men's toilet room when not in use from the north side of the site. A urinal is also provided. Lower left: An evening view of the exterior when in use, with opaque glass. Right: At night, the entire facility illuminates the park like a lantern.

井の頭通り側から見た、夜の「はるのおがわコミュニティパークトイレ」。便器などの衛生設備機器は、清掃がしやすい壁掛式を採用。床も拭き取りやすいエポキシ樹脂系塗床。

Evening view of the Haru-no-Ogawa Community Park Public Toilet from Inogashira Dori. The toilets and other sanitary fixtures are wall-hung for easy maintenance. The floors are coated with epoxy resin for easy wiping.

「代々木深町小公園トイレ」の断面イメージ。天井高は約2.1m。左からユニバーサル、女性用、男性用トイレ。プランは「はるのおがわコミュニティパークトイレ」も同じ。

Cross-sectional sketch of the Yoyogi Fukamachi Mini Park Public Toilet. The ceiling height is approximately 2.1m. From left to right: universal, women's, and men's toilet rooms. The floor plan is the same for the Haru-no-Ogawa Community Park Public Toilet.

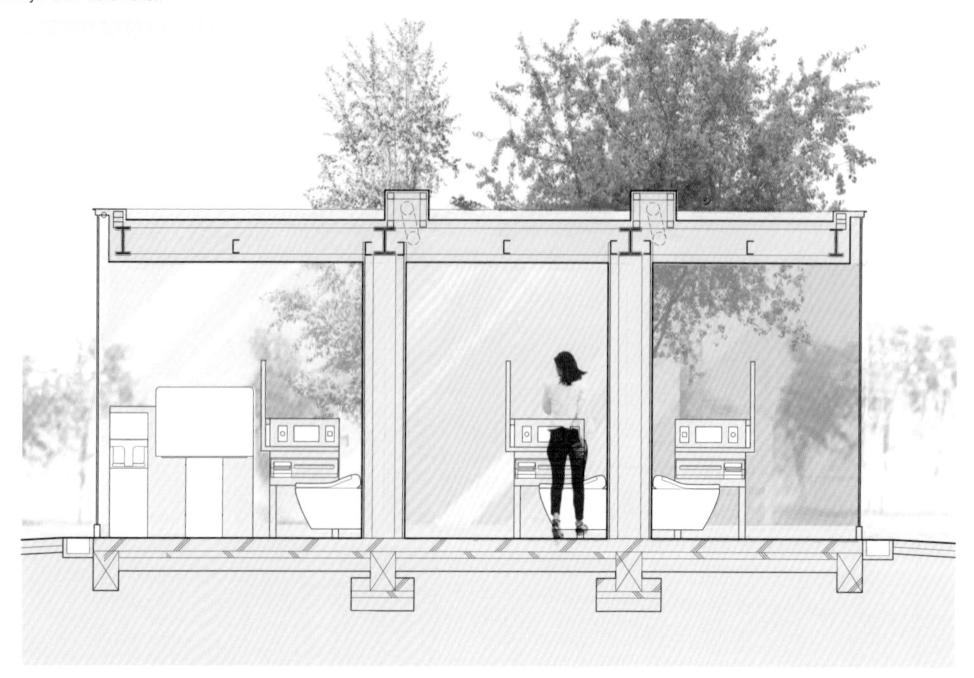

利用前の心配に応える、クリーンな透明トイレ

Clean, Transparent Toilet Facility That Addresses Pre-Use Concerns

公共のトイレ、特に公園にあるトイレは、入るときに2つの心配なことがあります。ひとつは中が綺麗（クリーン）かどうか、もうひとつは中に誰も隠れていないか。特に女性の利用者は、不安に思っている人が多いと思います。そのどちらをも解決するデザインとして、このガラスの「透明トイレ」は誕生しました。タイトルはそのままストレートに、「ザ トウメイ トウキョウ トイレット」。

電気を通すことによって透明になる「瞬間調光ガラス」の技術は、今回のトイレのために開発されたものではありません。強化ガラスにカラーフィルムと瞬間調光フィルムを圧着させたもので、建築を専門とする人たちの間では、以前から広く知られていた技術であり、オフィスの会議室などにもよく用いられています。私自身の作品では、たとえば、2019年に完成したスイスの〈スウォッチ・オメガ シテ・デュ・タン〉内の部屋の仕切りにも使用した経験があります。

デザインにあたり注意深く扱ったことは、未使用時、つまり、透明になった時に外から見える機器のレイアウトです。公園内の施設として、常にそこにあるわけですから、その景色の一部として、機器のレイアウトが美しく、整っている必要がありました。

「ザ トウメイ トウキョウ トイレット」は、井の頭通り沿いの「はるのおがわコミュニティパーク」と「代々木深町小公園」の2カ所に設置されています。仕組みや設備の構成は同じで、どちらも待合いスペースはなく、ユニバーサル・トイレ、女性用、男性用の個室が並びます。

色については、赤が女性、青が男性という固定概念にとらわれないように、と考え、「はるのおがわコミュニティパーク」を寒色系、「代々木深町小公園」を暖色系のトーンで揃え、昼の自然光の下でも明るい印象になることを目指しました。夜には、美しい行灯のように公園を照らします。

Public toilets, especially those in parks, pose two concerns when one enters them. One is whether it is clean inside, and the other is whether anyone is hiding inside. We believe many users, especially female users, are concerned about them. The glass "transparent toilet facility" was conceived as a design solution to these concerns. It is straightforwardly called "The Transparent Tokyo Toilet".

The "instant dimmable glass" technology, which turns glass transparent when electricity passes through, was not explicitly developed for this facility. This technology, consisting of colored film and instant dimmable film crimped onto tempered glass,

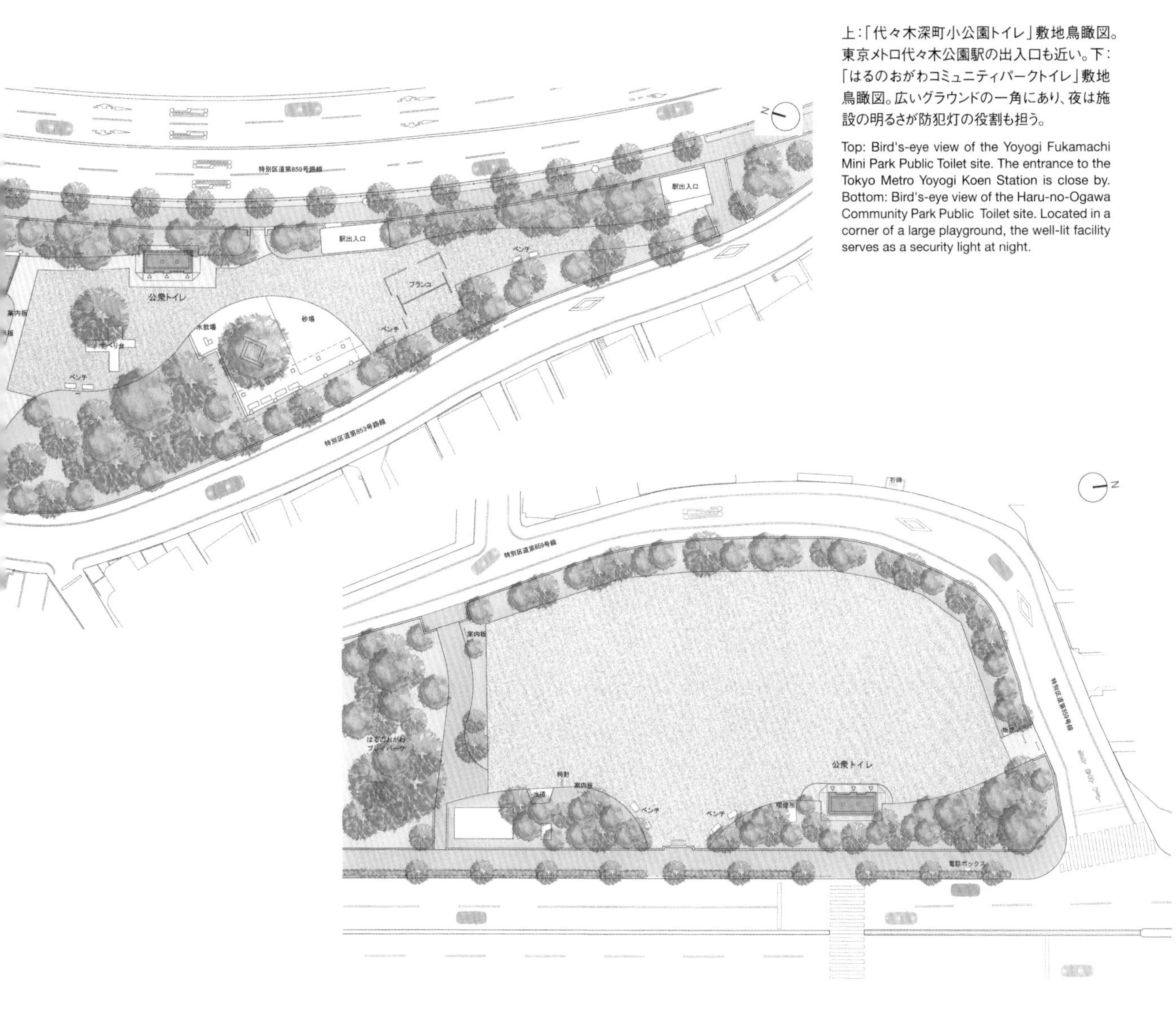

上:「代々木深町小公園トイレ」敷地鳥瞰図。東京メトロ代々木公園駅の出入口も近い。下:「はるのおがわコミュニティパークトイレ」敷地鳥瞰図。広いグラウンドの一角にあり、夜は施設の明るさが防犯灯の役割も担う。

Top: Bird's-eye view of the Yoyogi Fukamachi Mini Park Public Toilet site. The entrance to the Tokyo Metro Yoyogi Koen Station is close by. Bottom: Bird's-eye view of the Haru-no-Ogawa Community Park Public Toilet site. Located in a corner of a large playground, the well-lit facility serves as a security light at night.

has long been widely known among architectural professionals and is often used in office conference rooms. For example, we have used the same technology for the Swatch Omega Cité du Temps, completed in 2019 in Switzerland.

We paid careful attention to the equipment layout visible from the outside when not in use, meaning when the glass is transparent. Since it will always exist as a facility in the park, the equipment layout should be an aesthetically pleasing and well-organized part of the landscape.

The Transparent Tokyo Toilet is located in two locations along Inokashira Dori: Haru-no-ogawa Community Park and Yoyogi Fukamachi Mini Park. The structure and equipment layout of these facilities are the same: both have no waiting areas and are lined with universal, women's, and men's toilet rooms.

As for the colors, we wanted to avoid the stereotypes of red for female and blue for male. We chose cool tones for "Haru-no-Ogawa Community Park" and warm tones for "Yoyogi Fukamachi Mini Park" to create a bright impression under natural light in the daytime. At night, they illuminate the parks like beautiful lanterns.

Drawings and Specifications ▶P.242

坂 茂 Shigeru Ban

建築家。1957年生まれ。芝浦工業大学特別招聘教授。85年坂茂建築設計設立。95年、ボランタリー・アーキテクツ・ネットワーク(VAN)設立。紙管を利用した建築や避難所への間仕切りシステム提供など災害支援活動でも知られる。

Architect, born in 1957. Special Guest Professor, Shibaura Institute of Technology. Established Shigeru Ban Architects in 1985 and Voluntary Architects Network in 1996. He is known for his disaster relief activities, including the construction of paper tube architecture, and providing partitioning systems at evacuation centers.

恵比寿公園トイレ

EBISU PARK PUBLIC TOILET

デザイン：片山正通/Wonderwall® 渋谷区恵比寿西1丁目19番1号

Design : Masamichi Katayama / Wonderwall® 1-19-1 Ebisu-Nishi Shibuya-ku

2020年8月完成。壁式の鉄筋コンクリート造で、壁の最大横幅は約6.5m、高さ約3.8m。写真左の道路側に男性用、中央にユニバーサル、右に女性用トイレの入口がある。

Completed in August 2020. It is a walled reinforced concrete construction with a maximum wall width of about 6.5 m and a height of about 3.8 m. The entrance to the men's toilet is on the street side to the left of the photo, the universal toilet is in the center, and the women's toilet is on the right.

左上：男性用トイレの待合いスペースから
アプローチを見る。左下：女性用トイレ内観。
写真左の個室にはベビーチェアも備える。
右：公園北東の入口付近から見た外観。

Top left: View of the approach from the men's toilet waiting area. Bottom left: Interior view of the women's toilet. The cubicle on the left is equipped with a baby chair. Right: Exterior view from near the northeast entrance of the park.

男性用トイレ内観。個室は写真右に1室。小便器が2器。なお、女性用トイレの個室は2室。焼杉板の型枠の木目が残るコンクリート打ち放しの壁の仕上げは、個室内もすべて共通。

Interior view of the men's toilet. There is one cubicle to the right of the photo and two urinals. The women's toilet has two cubicles. The walls are finished entirely in concrete, with the wood grain of the burned cedar board formwork remaining, and the same finish is used in all the cubicles.

左：男性用トイレへのアプローチ。間接光は誘導灯でもある。右上：壁と壁の接点にはスリットを設け間接光を仕込んでいる。右下：壁の素材感が光によって浮かび上がる外観夕景。

Left: Approach to the men's toilet room. Indirect light also serves as a guide light. Top right: Indirect light is inserted in slits at the contact points between the walls. Bottom right: Evening view of the exterior, revealing the materiality of the walls under the light.

検討模型。壁の組み合わせによって生まれた多角形の空間の複雑さがよくわかる。

Study model. The complexity of the polygonal spaces created by combining the walls is clearly visible.

15枚の壁を組み合わせた、遊具のような現代の川屋

A Modern Kawaya Like a Piece of Playground Equipment Consisting of 15 Walls

「THE TOKYO TOILET」に参加することになって気づいたことのひとつは、自分自身、公共のトイレ、特に公園内のトイレはほとんど使ってこなかった、ということでした。たまたま使わずに済んできたのもありますが、残念ながら、使いたい施設だとは思ってこなかったのも、事実だと思います。そのため「恵比寿公園トイレ」では、この場所に対する公共トイレの「あり方」を捉え直し、利用者に「使いたい」と思ってもらえる場所にすることを目指しました。

店舗やオフィスなど、これまで手がけてきたインテリアのプロジェクトには、必ずトイレがあります。では、それらは何を拠り所にデザインをしているかというと、場所や空間のアイデンティティに寄り添うこと。その上で、単に「デザイン性」というより、使いやすさやプライバシーの両立を、常に慎重に考えてきました。

「恵比寿公園」周辺には、保育園や小学校があり、大人から子どもまで、さまざまな年代・性別の人が使う公園だということをまず念頭に置きました。以前恵比寿に事務所があったので、敷地周辺のことをよく知っていたのも構想の助けになりました。公園との一体感は強く意識していて、建築的なものから距離を置き、遊具やベンチや樹木のように何気なく公園に佇むオブジェクトとしてのあり方を探りたいと考えました。

コンセプトは「現代の川屋(厠)」。日本におけるトイレの起源は、川に直接用便する「川屋」と呼ばれるもので、縄文時代早期にまで遡ります。それは、土で固められたもの、木材を結び付けて作ったものなど極めてプリミティブで質素なものでした。そのため今回の建物を構成する素材も、原始的で質素なあり方につながるものを選びました。具体的には、コンクリート板に「木目」という自然の造形を施すことで、硬質な素材に素朴さや優しさを与えています。

15枚のコンクリート板を、感覚的には"無邪気に"組み合わせていますが、視線や動線に配慮した細かい計算を行っています。壁と壁の間は、男性用、女性用、ユニバーサル・トイレという3つの空間への導入に。人びとが不思議な遊具と戯れるようなユニークな関係性と、安心や安全の両立を図っています。壁と壁の接点に仕込んだ間接光は、このトイレをオブジェクトとして映し出す、演出の要。もちろん、機能的な照明として、利用者の役に立つものでもあります。

One of the things I realized when joining "THE TOKYO TOILET" project was that I had rarely used public toilet in the past, especially those in parks. It is not only because I never felt the need to use them, but also because, unfortunately, they were not the kind of facilities that I wanted to use. This is why the "Ebisu Park Public Toilet" sought to rethink the "ideal" public toilet facility for this location and make it a place that users would want to use.

Every interior design project we have worked on, be it a store or an office, has always involved toilet rooms. In designing

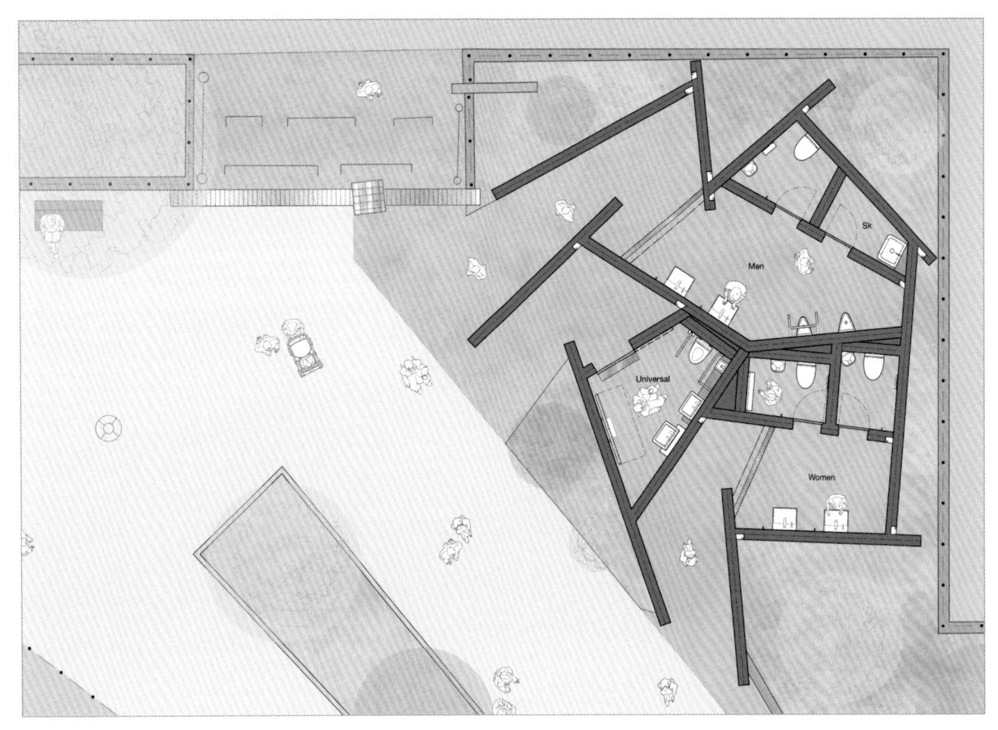

平面図。敷地面積198.41m²で、建築面積は37.52m²。既存樹木を活かして配置された。

Floor Plan. The site area is 198.41 m² and the building area is 37.52 m². The site plan takes advantage of the existing trees.

them, we stay true to the identity of the place or space. On this basis, we have always carefully sought to balance usability and privacy, rather than simply focusing on "design quality."

Since there are nursery schools and elementary schools in the vicinity of Ebisu Park, we gave first priority to the fact that the park is used by people of all ages and genders, from adults to children. Since we previously had an office in Ebisu, we are familiar with the area around the site, which was helpful in conceptualizing the project. A sense of integrity with the park was one of our main concerns, so we wanted to explore how to distance it from architectural elements and make it an object standing casually in the park, like playground equipment, benches, and trees.

The concept is a "modern kawaya." The origin of toilets in Japan dates back to the early Jomon period, when the structures called "kawaya" (literally a "river house": kawa means a "river" and "ya" means a "house") were built on rivers and people used them to relieve themselves directly upon them. They were extremely primitive and simple, made of earth or wood tied together. Accordingly, we chose primitive and simple materials to compose this building. Specifically, the natural formation of wood grain applied to the concrete slabs gives a sense of simplicity and gentleness to the hard material.

The fifteen concrete slabs are assembled in an intuitive and care-free manner , with detailed calculations of sight lines and flow lines. The wall-to-wall space is an entry into three spaces: men's, women's, and universal toilets. Our design seeks to balance safety and security while maintaining a unique relationship in which people interact with the mysterious playground equipment. The indirect light, positioned at the contact points between the walls, is the key to staging and illuminating the toilets as objects. It naturally serves as functional lighting for the users as well.

Drawings and Specifications ▶P.244

片山正通 Masamichi Katayama

インテリアデザイナー。ワンダーウォール代表、武蔵野美術大学空間演出デザイン学科教授。ファッションなどのブティックからブランディング・スペース、大型商業施設の全体計画まで、世界各国で多彩なプロジェクトを手がける。

Interior designer. Principal / founder of Wonderwall and professor at Musashino Art University's Department of Spatial Design. He undertakes a wide variety of projects worldwide, from fashion boutiques to branding spaces and overall planning for large-scale commercial facilities.

渋谷区東三丁目公衆便所
SHIBUYA HIGASHI 3 PUBLIC BATHROOM

東三丁目公衆トイレ

HIGASHI SANCHOME PUBLIC TOILET

デザイン：田村奈穂　渋谷区東3丁目27番1号

Design : Nao Tamura　3-27-1 Higashi Shibuya-ku

2020年8月完成。鉄筋コンクリート造。JRの線路沿いで、敷地面積は約37m²とわずかなスペース。写真左手からユニバーサル、男性用、女性用トイレが並ぶ。

Completed in August 2020. Reinforced concrete construction. Located along the JR railroad tracks, the site area is only about 37 m². Universal, men's, and women's toilet rooms are lined up from left to right in the photo.

渋谷区東三丁目公衆便所
SHIBUYA HIGASHI 3 PUBLIC BATHROOM

左：車椅子でも利用可能なユニバーサル・トイレの入口付近。右：敷地北側、三角形の土地の鋭角側からの外観。アプローチを斜めにとり、効率的に個室のスペースを確保している。

Left: View near the entrance of the wheelchair-accessible universal toilet room. Right: Exterior view from the acute-angled corner of the triangular lot on the north side. The approach is slanted to accommodate toilet cubicles efficiently.

上、下左、下右：ファサードの鉄板はできるだけ厚みを感じさせないよう20mmとした。下中：女性用トイレの個室。待合いスペースはとらず、左右に2室を振り分けている。

Top, bottom left, bottom right: The steel plates of the facade is only 20 mm thick to minimize perceived thickness as much as possible. Bottom middle: View of a cubicle in the women's toilet room. The women's toilet room is divided into two cubicles on either side without a waiting area.

敷地対面からの外観。ユニバーサルをはじめ3つのトイレを均等なバランスで配置。安全とプライバシーに配慮し、個室はアプローチからすぐの場所に設け、手洗い場も個室内に収めた。

Exterior view from across the street. The three toilet rooms, including the universal toilet room, are evenly spaced. Considering safety and privacy, the toilet cubicles are located right off the approach, and the hand wash basins are contained within.

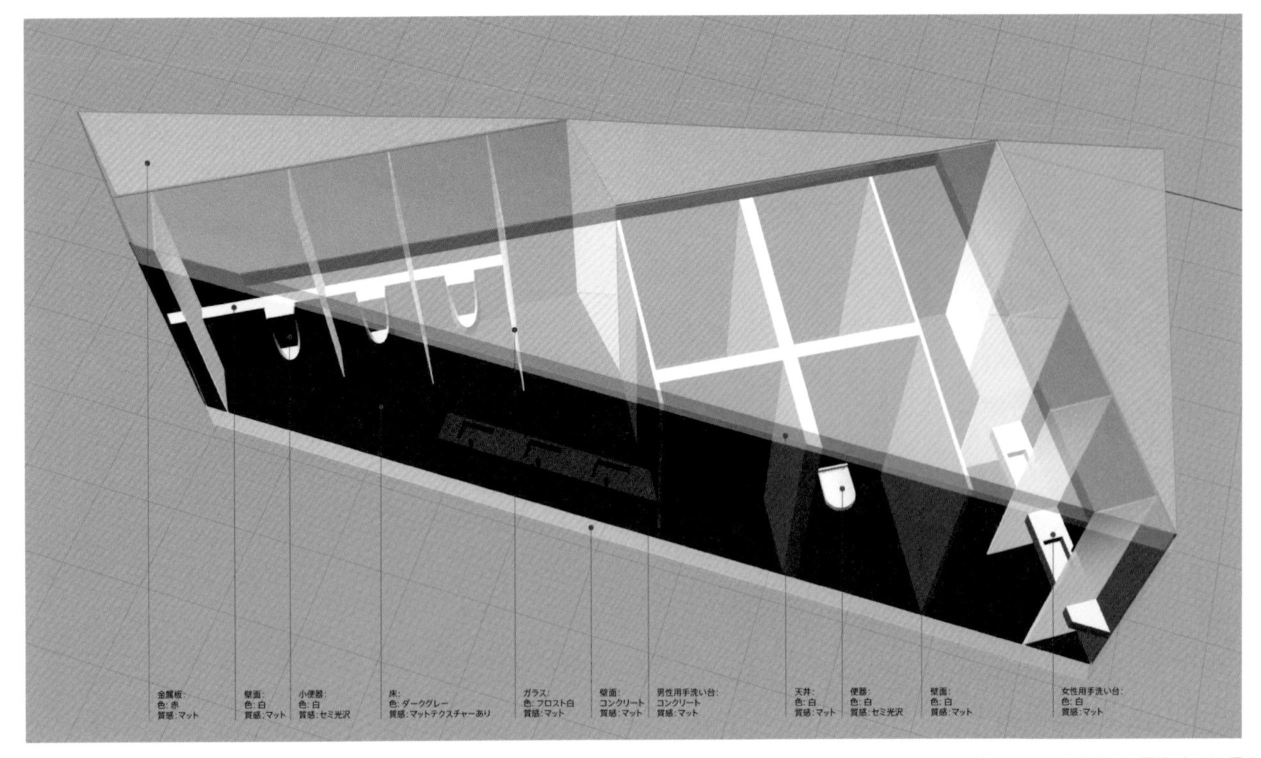

検討パース。最終的な配置場所などは異なるが、男性用トイレには個室1室に加え、小便器も3器備えられた。

Perspective view of the study plan. The men's toilet room contains one cubicle and three urinals, albeit not in the exact location as the final layout.

安全性を第一に考えた、アラートカラーの三角の箱

Alert-Colored Triangular Box Prioritizing Safety

「東三丁目公衆トイレ」をデザインするにあたり、まず思い浮かべたのは、私の住むニューヨークで出会った「LGBTQ+」の人びとの姿です。彼らは、自分の認識する性に正直に生きています。ありのままの「自分」を生き、その“ありのまま”を受け入れる社会。そのような社会を想像しながら、公共トイレを突き詰めて考えた結果、大切なのは、トイレを利用する“誰もが”同じように、快適な気持ちを得られるための「プライバシーと安全」であると思い至りました。「東三丁目公衆トイレ」ではそのことを念頭に、個人の空間を再定義し、3つの空間をデザインしました。

背景に鉄道との境になる大きなコンクリートの壁、左右に非常倉庫と電車の電源分配ボックス、交通量の多い車道に挟まれた、都会の真ん中に出来た小さな三角形の「隙間」。それが、この敷地を見たときの第一印象です。その小さな三角形の敷地に対して、日本の折り紙の原型でもある伝統作法、「折形」からインスピレーションをもらい、紙を折ることでつくられる幾何学的な形を空間化してみようと考えました。

紙を折り、贈り物を包むということは、清潔で丁寧で、その作法には「相手を尊ぶ気持ち」が込められていると言います。このトイレもまた、清潔で丁寧で、お互いを尊び合うような場所になってほしい。国際都市渋谷にやってくるビジターへのもてなしの気持ちと、利用する人びとを包み込む安全な場所にしたいという思いも込めています。

「折形」の空間化にあたっては、壁や天井の物理的な厚みを極力感じさせないよう配慮し、鉄板の薄さと紙を折ったような表現にこだわりました。鉄板を使うことに難しさはありませんでしたが、鉄の断熱塗料が白みがかった色をしていて、外壁の色を、今回実現させたかった「鮮やかな赤」に近づけるのに苦労しました。

「赤」は、煩雑な立地環境において、ここにトイレがあることを明確に伝える色であり、心理的な緊張感を持たせる「アラートカラー」でもあります。夜間やひとけのない時に公共トイレを利用する際、恐怖心、あるいは緊張感を覚えたことのある方は少なくないと思います。残念ながら、トイレが犯罪の現場になってしまうこともあります。この「赤」が衝動的な犯罪を抑止し、利用者が安心して使える公共トイレであり続けることを願っています。

The first thing that came to mind when designing the "Higashi Sanchome Public Toilet" was the LGBTQ+ people I met in New York City, where I live. They lead their lives with candor, accepting their sexuality in a society that allows people to live their lives as they are and accepts them for who they are. Imagining such a society, we took thorough considerations into what would be the most important factors in designing a public toilet facility and concluded that privacy and safety are key in ensuring the same comfort for everyone using the toilets. With this in mind, we redefined how a personal space should be and

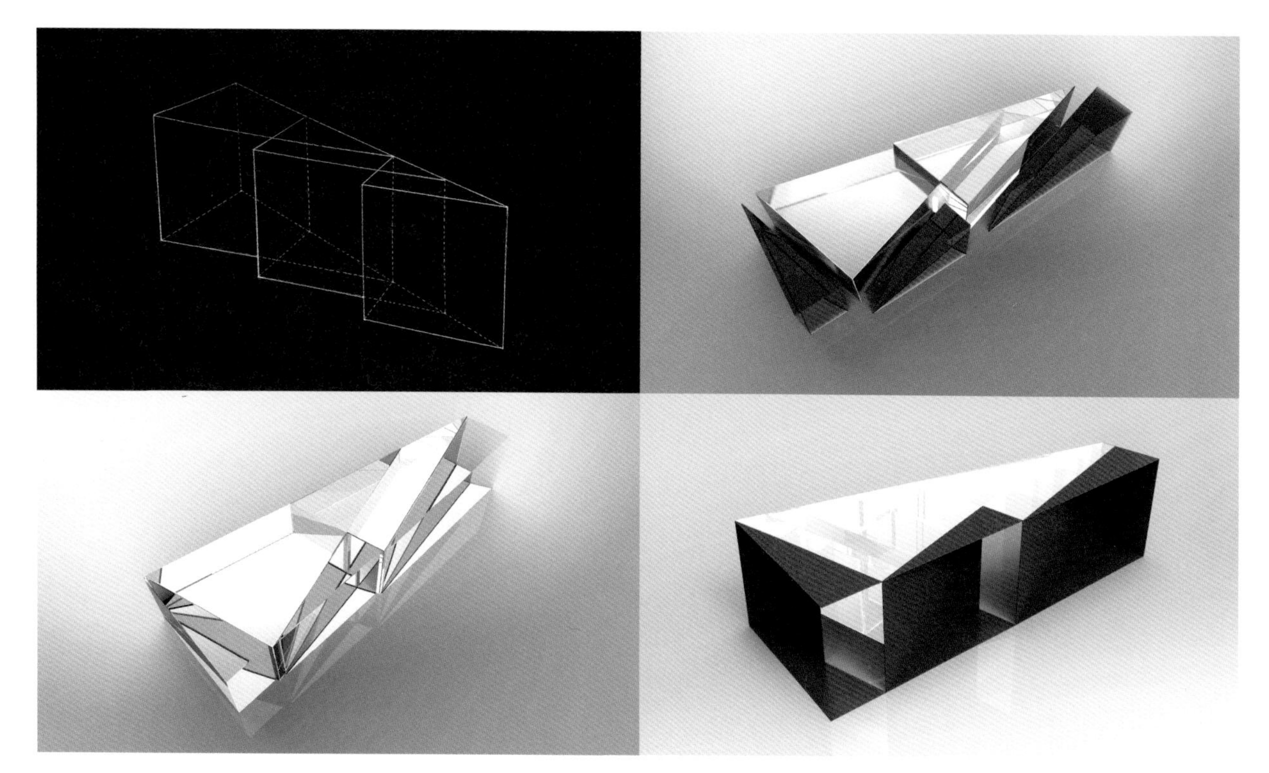

ダイアグラム。細長い三角形の敷地に対しての均等なゾーン設定と、「折形」をイメージした造形の思考過程がわかる。

Diagram showing the thinking process involving the even zoning on the elongated triangular site and form-making inspired by origata.

designed three spaces for the Higashi Sanchome Public Toilet.

Upon seeing the site, our first impression was a small triangular gap in the middle of the city, sandwiched between a large concrete wall bordering the railroads in the background, an emergency warehouse and train power distribution box on either side, and a busy street. For this small triangular site, the traditional art of origata, an original form of Japanese origami, inspired the idea of spatializing the geometric shapes made by folding paper.

Folding paper and wrapping gifts suggest cleanliness and courteousness, and the gesture is imbued with a sense of respect for the other person. We wanted this toilet facility to be a place that is also clean, courteous, and respectful of each other and make it a welcoming place for visitors to the cosmopolitan city of Shibuya and a safe place embracing the users.

In transforming origata into space, we took care to minimize the physical thickness of the walls and ceiling and focused on expressing the thinness and the folded paper-like appearance of the steel plates. While using steel plates was not particularly challenging, the insulating paint on the steel had a whitish color, and we struggled to bring the color of the exterior wall closer to the vivid red that we wanted to achieve for this project.

Red is a color that clearly communicates the presence of a toilet facility in the complex location and serves as an alert color that evokes a sense of tension. Many of us have experienced fear or nervousness when using public toilets at night or when no one is around. Unfortunately, public toilet facilities often turn into crime scenes. It is our hope that this red color will help to deter impulsive crimes and continue to be a public toilet facility that users can use with a peace of mind.

MAP 13

Drawings and Specifications ▶ P.246

田村奈穂 Nao Tamura

プロダクトデザイナー。工業デザインを専門にするアメリカのデザイン会社、スマート・デザインを経てnownao inc. 設立。ニューヨークを拠点に活動を続けている。コミュニケーションデザインを軸に、時計からインスタレーションまで幅広く活躍。

Product designer based in New York. Founded nownao inc. after working for Smart Design, a U.S. design company specializing in industrial design. Her work ranges from clocks to installations, with a focus on communication design.

恵比寿東公園トイレ

EBISU EAST PARK PUBLIC TOILET

デザイン：槇文彦　渋谷区恵比寿1丁目2番16号

Design : Fumihiko Maki　1-2-16 Ebisu Shibuya-ku

2020年7月完成。鉄筋コンクリート造、一部鉄骨造。分散配置されたユニバーサル、男性用、女性用トイレを軽やかな屋根がつなぐ。写真中央のボックスがユニバーサル・トイレ。

Completed in July 2020. Reinforced concrete construction with additional steel elements. A light roof connects the dispersed universal, men's, and women's toilet rooms. At the center of the photo is the universal toilet room.

左上：敷地南側の歩道からの外観。左下：屋根の鉄板の厚みは16mm。風通しと採光のために中央を開けた。右：ユニバーサル・トイレ北側のベンチは建物と一体でデザインされている。

Top left: Exterior view from the sidewalk on the south side of the site. Bottom left: The thickness of the steel plate roof is 16mm and is open in the center for natural ventilation and lighting. Right: The bench on the north side of the universal toilet room is designed as an integral part of the building.

左：分散配置により通り抜け可能で防犯性も高い。屋根下の高窓からの採光で個室内も明るい。右上：ユニバーサル・トイレ入口付近から中庭を見通す。右下：中庭の樹木はカツラ。

Left: The dispersed layout allows people to pass through and ensures high security. The toilet cubicles are brightly lit by natural light from the high windows under the roof. Top right: View towards the courtyard from the entrance to the universal toilet. Bottom right: A katsura tree (Japanese Judas tree) defines the courtyard.

公園内からの遠景。恵比寿東公園は1958年開園の歴史ある児童遊園で、乳幼児連れも多く、男性用、女性用トイレともに個室内にベビーチェアとベビーシートを配置している。

Distant view from inside the park. Ebisu East Park is a historic children's playground that opened in 1958. Since many people bring their infants and toddlers to the park, baby chairs and baby seats are provided in both men's and women's toilets.

屋根の鳥瞰パース。中央を開け、いくつものカーブを組み合わせた全体像がわかる。ふわりと布をかぶせたかのように軽やかで美しい。

Bird's-eye view of the roof with an opening in the center and a combination of several curved planes. The roof is light and beautiful, as if softly covered with fabric.

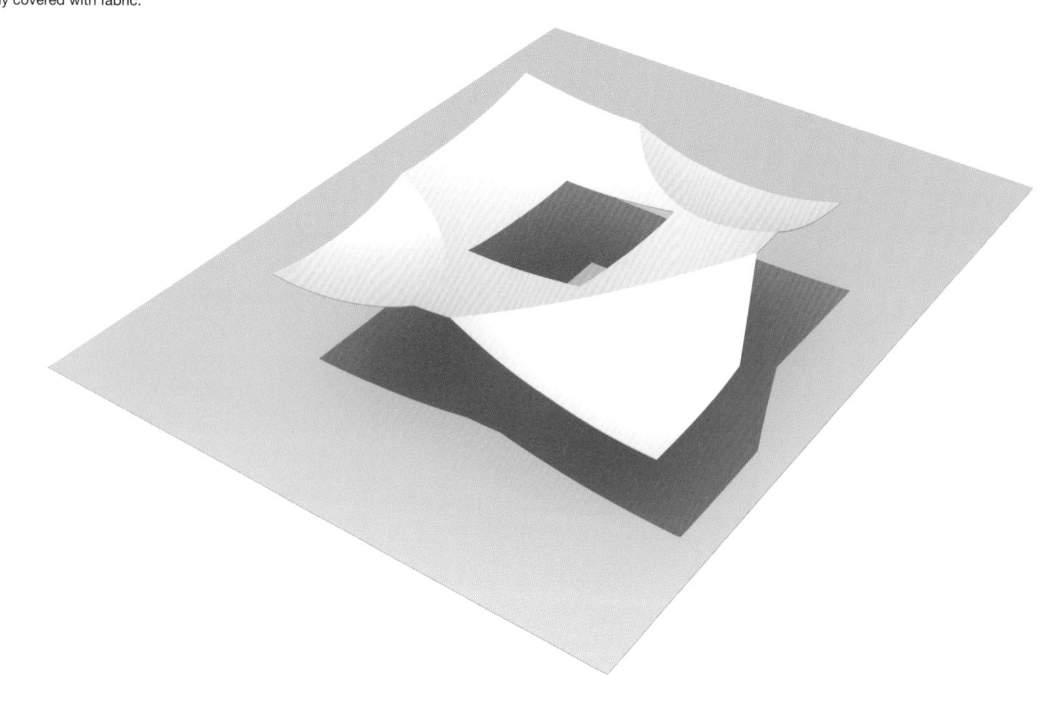

中庭とベンチを備えた、公園内の爽やかなパビリオン

Refreshing Pavilion in the Park with a Courtyard and a Bench

私が設計において大切にしてきたことのひとつに「場所性」があります。その場所の歴史的背景や普遍的な価値といったことです。

敷地の恵比寿東公園は、緑豊かな児童遊園として普段から近隣の人びとに親しまれている公園です。そのため、今回の「恵比寿東公園トイレ」では、ここが長く愛されてきた児童遊園の中だからこそ、爽やかで、少しユーモアのある建築がいいのではないかと考えました。爽やかというのは、開放的で明るく、清潔な環境であり、ユーモアというのは、使う人の緊張をほぐし、リラックスできる場所づくりということでもあります。

子どもたちから通勤中の人びとまで、多様な利用者に配慮し、具体的には、施設ボリュームの分散配置によって視線を制御しながら、安全で快適な空間の創出を目指しました。どの方角からもアプローチ可能で、この建物には死角となる「裏」がありません。ボリュームを統合する軽快な屋根は、通風を促し、自然光を呼び込む形態としています。また、中央には、緑豊かな公園と一体になる中庭を、ユニバーサル・トイレの外側には腰を下ろしてひと息つけるベンチも備え付けました。

私たちは計画当初から、この施設を単なるパブリックトイレとしてだけでなく、休憩所を備えた「公園内のパビリオン」として機能する公共空間としたいと考えていました。ベンチはトイレの利用者だけではなく、公園の利用者にもぜひ使っていただきたいと思っています。

建物全体から受ける印象としては、公園らしく、遊具のような「ユニークな姿」によって、設計当初に意図した「ユーモア」を感じていただくことができるのではないかと考えました。恵比寿東公園はタコの遊具によって「タコ公園」とも呼ばれています。屋根の形状からこの建築を「イカ」に見立てられることは、むしろ大歓迎で、タコ公園に新しく生まれた「イカのトイレ」として親しまれることを望んでいます。

場所を活かし、皆さんによろこんでもらえる公共空間のあり方について、「恵比寿東公園トイレ」のような考え方もあるのだと知ってもらうことも、大切なことだと思います。

建築は永い時間、その場所にあり続けます。だからこそ長期的な視点で、社会の財産になるものであってほしい。ディーセント(decent：品格のある)、年月を経ても恥ずかしくないものであること。「恵比寿東公園トイレ」ではその思いを、新たな形で展開できたのではないかと思っています。

One of the things I have always focused on in my design work is the notion of "placeness," which refers to the historical background and universal value of a place.

The site, Ebisu Higashi Park, is a lush green children's playground frequented by its neighbors. Precisely because this is a well-used children's playground, we thought a refreshing and somewhat humorous architecture would be appropriate. "Refreshing" means an open, bright, and clean environment,

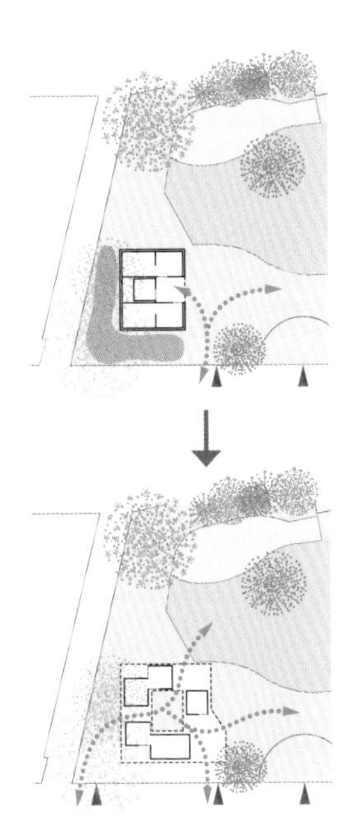

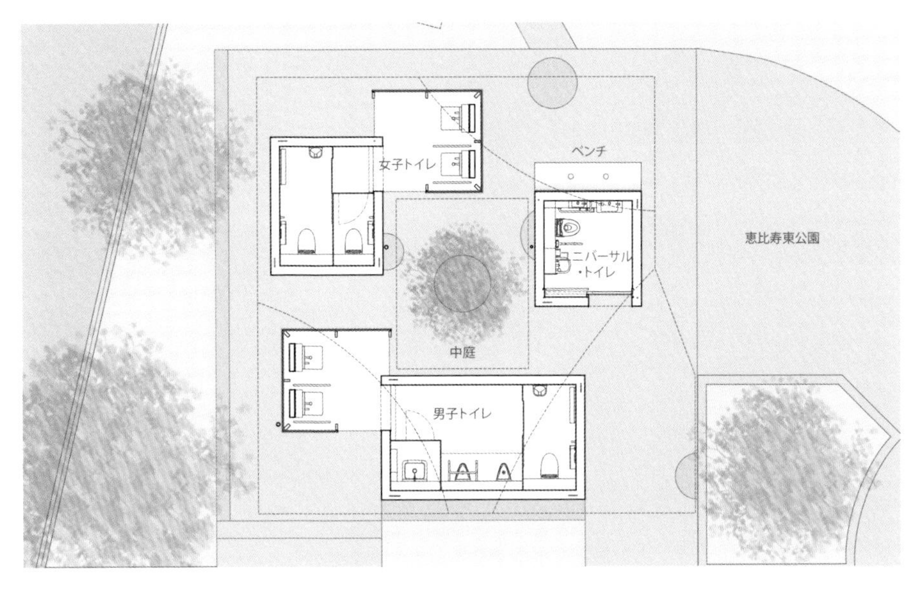

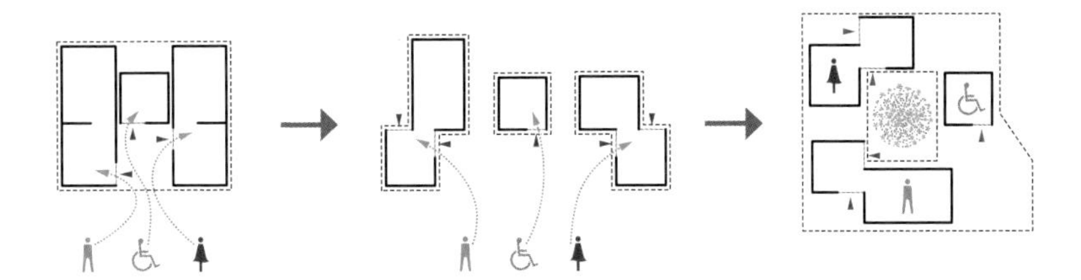

左から時計回りに、施設の「裏」を作らないことを意図した動線図、平面図、分散配置のダイアグラム図。敷地面積は約215m²。

Clockwise from left, a circulation diagram that shows how the design avoids "back" sides, plan, and diagram of the dispersed layout. The site area is about 215m².

and "humorous" means creating a place where people can unwind and relax.

Taking into consideration the diverse users, from children to commuters, we specifically sought to create safe and hospitable spaces while controlling sight lines via the dispersed placement of building volumes. Accessible from any direction, these buildings have no blind areas that create hidden spots. The light roof that brings together the volumes is shaped to facilitate ventilation and draw in natural light. In the center of the building is a courtyard blending in with the lush green park, and a bench is provided outside the universal toilet room for people to sit down and rest.

From the outset of our planning, we wanted this facility to be more than just a public toilet but a public space that would serve as a "pavilion in the park" with a rest area. The bench is intended for the use of park visitors as well as the toilet users.

As for the overall impression of the architecture, we thought that a design reminiscent of playground equipment would convey the sense of "humor" we had envisioned early on in the design process. Ebisu East Park is nicknamed "Octopus Park" for its octopus playground equipment. The roof shape makes this architecture look like a squid. We welcome this association, and hope the toilet becomes known to visitors as the "Squid Toilet."

Creating public space that everyone can use and enjoy is essential for architecture. And it is important that architecture maintain this fundamental decency for many years. We hope that Ebisu East Park Public Toilet will exhibit these qualities, even as it explores new and interesting formal qualities.

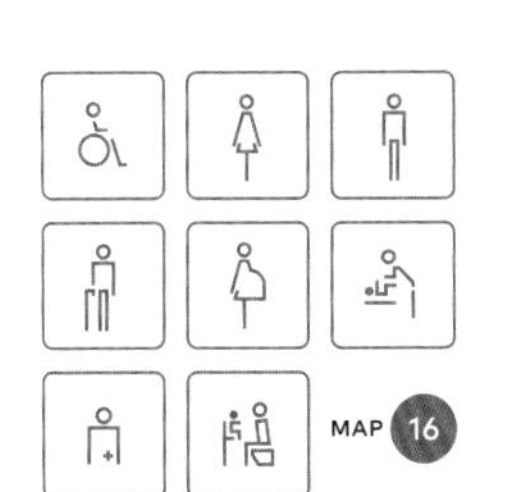

Drawings and Specifications ▶P.248

槇文彦 Fumihiko Maki

建築家。1928年生まれ。65年に槇総合計画事務所設立。ハーバード大学や東京大学で後進の指導に力を注ぐなど現代の建築界を牽引してきた建築家のひとり。代官山の街並みを形成した「ヒルサイドテラス」などの作品は世界的に高い評価を得ている。

Architect, born in 1928. Established Maki and Associates in 1965. He is widely recognized as one of the leading architects today and has devoted himself to teaching younger generations of architects at Harvard University and The University of Tokyo. His works, including the Hillside Terrace, which shaped the streetscape of Daikanyama, are highly acclaimed worldwide.

2020年8月完成。鉄筋コンクリート造、一部鉄骨造。京王線の地下化に伴って生まれた、緑道沿いの細長い敷地で、中央にユニバーサル、両端に男女共用トイレを配した。

Completed in August 2020. Reinforced concrete construction and partially steel frame construction. The long, narrow site along the greenway, resulting from the undergrounding of the Keio Line, has a universal toilet room in the center and gender-free toilet rooms at both ends.

西原一丁目公園トイレ

NISHIHARA ITCHOME PARK PUBLIC TOILET

デザイン：坂倉竹之助　渋谷区西原1丁目29番1号

Design : Takenosuke Sakakura　1-29-1 Nishihara Shibuya-ku

左：施設西側の側面。右上：外壁は青みがかった乳白色のガラス壁。右下：建替えと同時に施設前面の歩道も整備された。車椅子やベビーカーなどでもアプローチしやすい。

Left: View of the west side of the facility. Top right: The exterior walls are made of bluish milky-white glass. Bottom right: The sidewalk in front of the facility was renovated as part of the reconstruction and is easily accessible for wheelchairs and baby strollers.

照明が灯ると、外壁のガラス壁にも樹木のシルエットが浮かび上がる。左右の男女共用トイレは手すりなどの配置も左右対称とし、障がいなどに応じて使いやすいほうを選べるようにした。

When the lights turn on, the silhouettes of the trees appear on the exterior glass walls. The gender-free toilet rooms on either side of the facility are symmetrically arranged together with the handrails so that people with disabilities can choose the one they prefer to use.

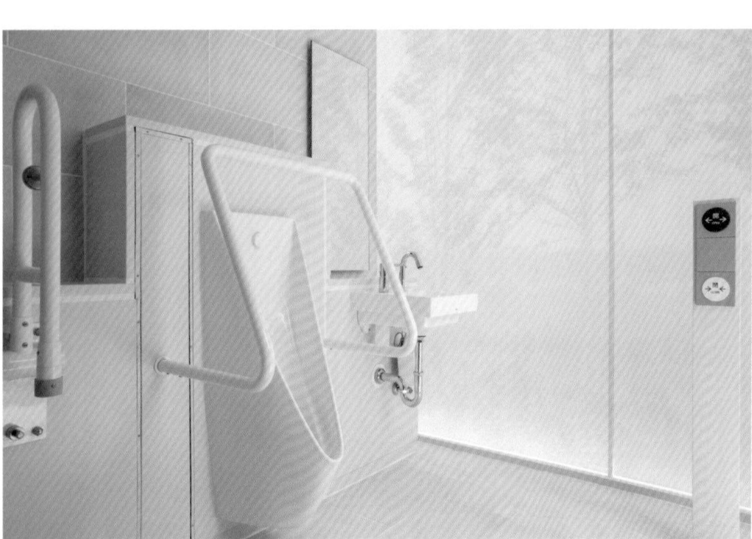

左上、左下：男女共用トイレは2辺がガラス壁。右：扉は自動スライド式。個室内の手洗い場も自動水栓とするなど、衛生面からできるだけ非接触で利用できる設備設計を行った。

Top left, bottom left: The gender-free toilet rooms have glass walls on two sides. Right: The door is an automatic sliding type. The facilities were designed to allow non-contact operation as much as possible from a hygienic standpoint, including automatic faucets for hand wash basins in the rooms.

歩道からの夕景。周辺は住宅街。日常的に敷地前を通る住民にとって、「行燈」の明るさが安心につながる。敷地面積は約226m²。

Evening view from the sidewalk. The neighborhood is residential. The brightness of andon (lantern) provides a sense of security for the residents passing by the site daily. The site area is approximately 226m².

公園と歩道を明るく照らす、端正な白い行燈
A Neat White "Andon" Brightly Illuminating the Park and the Sidewalk

公園を明るく照らす、「行燈」としての公共トイレをデザインしました。そのデザインの原点には、建替え前のトイレの利用頻度が極めて少なく、暗く近寄りがたいものであったこと、「西原一丁目公園」全体に寂しげな印象があったことなど、実際に敷地を訪れて感じたこの場所特有のさまざまな立地環境がありました。

このトイレを、多くの人びとに役立つ「行燈」にするために心がけたのは、建物の存在を意識させず、できるだけ明るさを確保することです。長方形のシンプルな形状は、その目的に対する答えとしてあります。

もうひとつは、トイレ内部を「気持ちよく使える」空間にすること。建築家として、これまで多くの住宅を手がけ、同時に多くのトイレを設計してきました。広さが限られているとはいえ、日々の生活に必要不可欠な場所であるトイレが、あまりにも閉鎖的で、気持ちよく過ごせる場所ではない、というのは望ましくないことと常々思ってきました。

そのため、今回「西原一丁目公園トイレ」をデザインするにあたっては、限られた敷地の中で、できるだけ明るく、開放的で気持ちのいい場所にしたい、という思いが強くありました。天井高を高くとり、透過性のあるガラスの外壁に森の景色を転写したのはそのためです。のびやかな空間にまるで木漏れ日が差し込んでくるかのような、清々しい室内になっていると思います。建物前面の通路に植樹をした木々がやがて大きく育ち、より気持ちのいい場所にしてくれることを期待しています。

個室はユニバーサル・トイレも含め、3室とも男女共用としました。公園自体が大きくはないこともありますが、一般的なトイレ整備の目標とされる便器数の充足、待ち時間の解消といった“数”や“時間”の数値とはまた別の魅力を持たせることで、みんなが「利用したい」と思うトイレを創出することが、この敷地では重要だと考えました。

待合いスペースもありません。これは、デッドスペースを作らないという防犯性を高める目的と、できるだけ余計なものを切り落としていくという方針からです。

「THE TOKYO TOILET」に共感した理由は、プロジェクト発足当初からメンテナンスを非常に重要視されていたことです。きれいなものをつくっても維持できないのでは意味がない。余計なものを切り落としたシンプルな形状は、メンテナンスのしやすさへの配慮でもあります。

We designed a public toilet as an "andon" (lantern) brightly illuminating the park. The starting point for the design was the site's conditions: the toilet facility, dark and inaccessible, were rarely used and the Nishihawa Itchome Park itself seemed deserted.

To make this facility a useful " lantern" for many people, we tried to make it as bright as possible without drawing attention to the presence of the building. The simple rectangular shape was the answer to this purpose.

Another important point was to make the interior of the toilet

男女共用トイレは2辺がガラス壁。時間帯によっては、ガラスに転写した樹木のシルエットと、施設周囲の植栽からの木漏れ日が重なる。

The men's and women's toilet rooms have glass walls on two sides. Depending on the time of the day, tree silhouettes cast onto the glass surface overlap with the sunlight filtering through the trees around the facility.

a "comfortable and user-friendly" space. We have worked on many houses and designed many toilet rooms in our architectural career and believe that it is undesirable to have a toilet room, an essential place for daily living, that is too closed off and unpleasant to spend time in, even though its size is limited.

For this reason, our design for the Nishihara Itchome Park Public Toilet was strongly motivated by our desire to make the site as bright, open, and pleasant as possible within the limited space available. To this end, the ceiling height was raised, and the forest scenery was transcribed onto the permeable glass exterior walls. The interior is refreshing, as if sunlight is streaming through the trees into the spacious space. We expect that the trees planted along the pathway in front of the building will soon grow and make the place even more pleasant.

All three toilet rooms, including the universal toilet room, are gender-free. This is partly because the park itself is relatively small, but also more importantly because we thought it was important to create a toilet facility that everyone would want to use, not only by satisfying the number of toilets and eliminating the waiting time, which are generally considered goals for designing toilet facilities, but by adding attractive qualities to the site.

The facility has no waiting area based on our strategy for enhancing security by eliminating dead space and cutting down on unnecessary things as much as possible.

We felt great sympathy for THE TOKYO TOILET because the project emphasized the importance of maintenance from the beginning. There is no point in creating something beautiful if it cannot be maintained. The simple shape of the facility, with all unnecessary things trimmed off, is intended for ease of maintenance.

坂倉竹之助 Takenosuke Sakakura

建築家。1946年生まれ。70年に坂倉建築研究所入所。79年現・坂倉アトリエ設立。坂倉建築研究所代表取締役会長。父・坂倉準三の意志を継ぎモダニズムの精神と「人間に根ざした建築」を信条としている。作品に「東京ミッドタウン」住居棟など。

Architect, born in in 1946. Takenosuke Sakakura joined Sakakura Associates in 1970 and established the present Sakakura Atelier in 1979. He is Representative Director and Chairman of the Sakakura Associates architects and engineers. Following in the footsteps of his father, Junzo Sakakura, he believes in the spirit of modernism and "architecture rooted in humanity. His works include the Tokyo Midtown Residences.

Drawings and Specifications ▶P.250

2020年9月完成。鉄筋コンクリート造、一部鉄骨造。個室を収めた円形の棟の周囲に格子をめぐらせ、その間を通路とした。出入口は2カ所あり、通り抜けられるようになっている。

Completed in September 2020. Reinforced concrete construction, partly steel construction. The circular facility containing the toilet rooms is surrounded by latticework, with a passageway between them. Two points of entry and exit allow people to pass through.

神宮通公園トイレ

JINGU-DORI PARK PUBLIC TOILET

デザイン：安藤忠雄　渋谷区神宮前6丁目22番8号

Design : Tadao Ando　6-22-8 Jingumae Shibuya-ku

左：大きく迫り出した庇の長さは最大で約3m。右上：公園内の大きな樹木はサクラ。明治通り沿いで、すぐ側にバス停もある。右下：敷地の南側正面にユニバーサル、円形の壁に沿って右が女性用、左が男性用トイレ。

Left: The maximum depth of the overhanging eaves is about 3m. Top right: Large trees in the park are cherry trees. The facility is located along Meiji Dori, with a bus stop right next to it. Bottom right: The universal toilet room faces the south front of the site, with the women's toilet room on the right and the men's toilet room on the left of the circular wall.

手洗い場は男女ともに、通路の途中にある。通路は通り抜け可能とすることで安全性を高めた。外周の格子も中央の棟の外壁も素材は耐久性の高いアルミを使用している。

Hand wash basins for the men's and women's toilet rooms are located in the passageway. The passageway allows passage through to enhance safety. Both the latticework around the perimeter and the exterior wall of the facility are made of aluminum, a highly durable material.

左上：男女ともに個室は1室。ベビーチェアも双方に設置した。左下：ユニバーサル・トイレ内。右：女性用トイレの個室への通路。アルミの縦格子から光が差し込み、風も抜ける。

Top left: There is one cubicle for both the men's and women's toilet rooms. Baby seats are installed in both. Bottom left: Interior view of the universal toilet room. Right: A passageway to a cubicle in the women's toilet room. Light and air filter through the vertical aluminum latticework.

安藤忠雄によるスケッチ。円形平面の建物と長い庇という建築の主軸が描かれている。
Sketch by Tadao Ando depicting the key architectural concept of the facility with a circular floor plan and deep eaves.

円形の建物から長い庇が迫り出した、「雨やどり」トイレ
A "Rain Shelter" Toilet Facility with Deep Eaves Protruding from the Circular Building

世界に誇る日本の良さは、美しくて清潔で礼儀正しいことです。日本のトイレは、その良さを象徴する場所のひとつではないでしょうか。これまで世界中を旅してきた経験を振り返っても、日本のトイレほど、美しく清潔で、あとに使う人への礼儀を感じるトイレはありません。

美しくて清潔であることに関して大事なことは、風通しです。湿度の高い日本では、昔から建物は風通しに配慮して設計されてきました。住宅でも寺院でも、伝統的な日本建築に深い庇と縁側があるのも、風通しを考えてのことでした。一方で、現代の感染症予防の鍵のひとつも、風通しと言われています。過去を見ても、これからの社会を考えても、明るく、風通しがよく、そして安全なトイレというのが、公共トイレのあるべき姿ではないかと考えます。

今回の「神宮通公園トイレ」では、円形の建物の外壁を縦格子とし、建物内部に光と風が通るようになっています。円というのは、形の原点であり、誰もが使える開かれた施設の形としても、ふさわしいのではないかと思います。

トイレのあり方として最初に考えたことは、小さな“あずまや”なりに、公共トイレという機能だけではない、都市施設としての意味、「パブリックな価値を持つもの」でありたい、ということです。屋根の庇を大きく迫り出し、雨が降ったら庇の下に逃げ込める、「雨やどり」ができるトイレをつくろう、と考えました。

これまで、数多くの建築を手がけてきましたが、私自身、トイレだけを単体で設計するのは、今回が初めてです。トイレが必要になったときはもちろん、ここが、困ったときに駆け込める「みんなの場所」として愛される建物であってほしいと願っています。

One of the best qualities of Japan that we can proudly present to the world is its beauty, cleanliness, and courteousness. The Japanese toilets are one of the places epitomizing this quality. Reflecting on my travels around the world, I have yet to find a toilet as beautiful, clean, and courteous to subsequent users as the ones in Japan.

The key to retaining its beauty and cleanliness is ventilation. In the humid country of Japan, buildings have always been designed to provide good ventilation. The deep eaves and engawa (a covered wooden corridor or veranda) in traditional Japanese architecture, including houses and temples, were also designed with ventilation in mind. On the other hand,

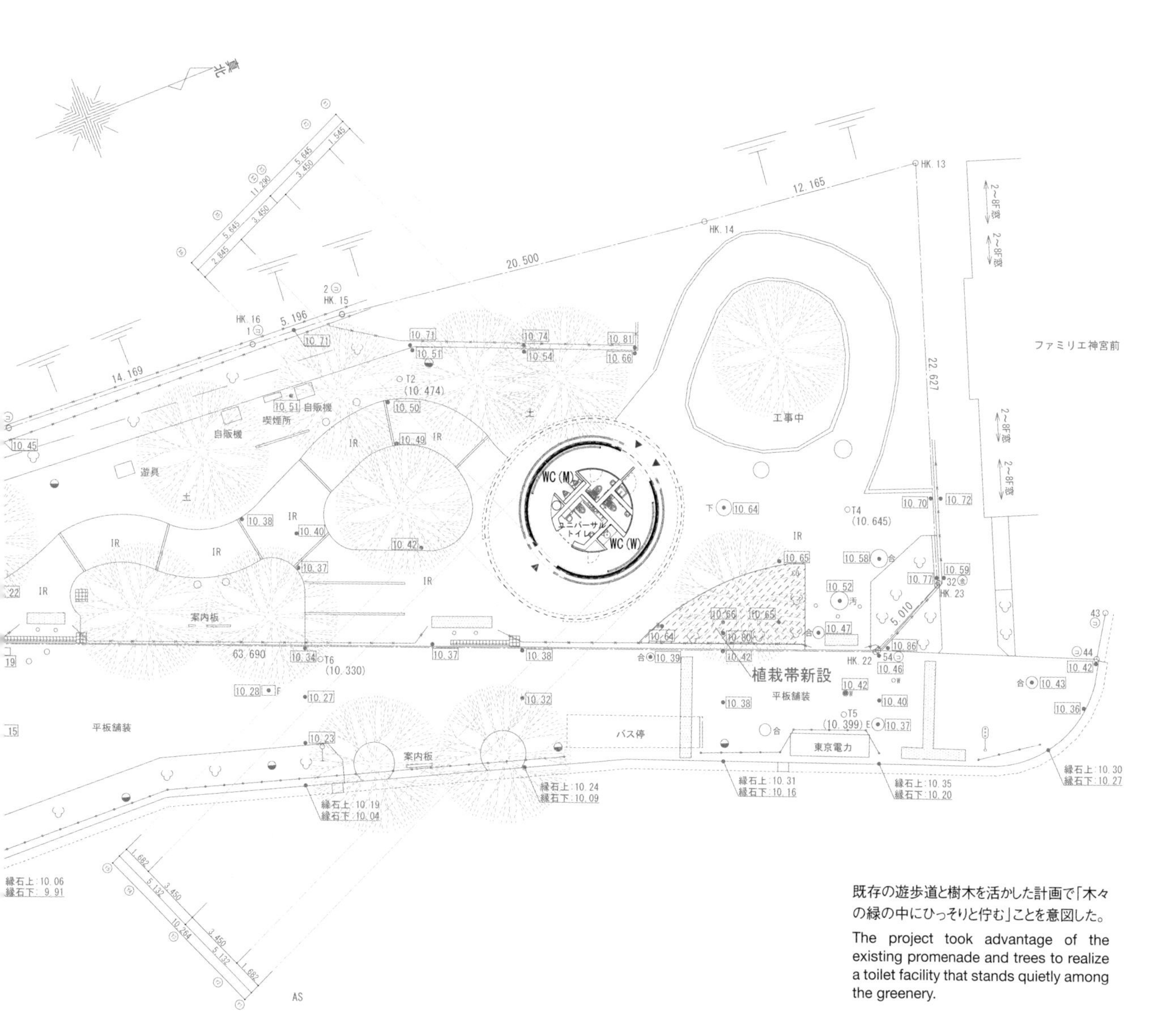

既存の遊歩道と樹木を活かした計画で「木々の緑の中にひっそりと佇む」ことを意図した。

The project took advantage of the existing promenade and trees to realize a toilet facility that stands quietly among the greenery.

ventilation is one of the keys to preventing the spread of infectious diseases today. Looking back at the past and forward to the future, public toilets should be bright, well-ventilated, and safe.

In this "Jingu-Dori Park Public Toilet," the exterior wall of the circular building is made of vertical latticework, allowing light and air to pass through inside the facility. The circle is the origin of shapes and is also appropriate as a form of an open facility for everyone to use.

Our initial idea for the facility was to create a small "azumaya" or a pavilion that would have a public value as an urban facility rather than simply serving the function of a public toilet. The eaves protrude out far, creating a rain shelter toilet pavilion where people can run under the eaves when it starts to rain.

I have undertaken many architectural projects, but this is my first time designing a toilet as a stand-alone structure. I hope people will cherish this facility as a place for everyone that they can run to in times of trouble, not to mention when they need to use it.

安藤忠雄 Tadao Ando

建築家。1941年生まれ。独学で建築を学び、69年安藤忠雄建築研究所設立。大阪を拠点に世界で活躍し、住宅から公共施設まで国内外に作品多数。植樹運動や自費を投じての児童向け図書施設の創設など、社会貢献活動にも精力的に取り組んでいる。

Architect, born in 1941. Self-taught in architecture, Tadao Ando established Tadao Ando Architect and Associates in 1969. Based in Osaka, he works internationally and has designed numerous works in Japan and abroad, ranging from residences to public facilities. He is also active in social philanthropy, including tree-planting campaigns and the establishment of children's library facilities using his own funds.

MAP 11

Drawings and Specifications ▶ P.252

2021年5月完成。鉄骨造、鉄筋コンクリート造。中央がユニバーサル・トイレ。左に男性用、右に女性用トイレ。高層のオフィスビルが立ち並ぶ、明治通り沿いの一角にある。

Completed in May 2021. Steel frame and reinforced concrete construction. The universal toilet room is in the center, the men's toilet room on the left, and the women's toilet room on the right. Located in a corner along Meiji Dori, lined with high-rise office buildings.

神宮前公衆トイレ

JINGUMAE PUBLIC TOILET

デザイン：NIGO®　渋谷区神宮前1丁目3番14号

Design : NIGO®　1-3-14 Jingumae Shibuya-ku

左上、左下：前面の歩道に対して水平に、間口を広く、奥行きの浅い横長の建物とした。右：敷地の囲いも一軒家の親しみやすさを意識し、木製のガーデンフェンスに。

Top left, bottom left: The building is long and shallow, with a wide frontage and level with the front sidewalk. Right: The site is enclosed with wooden garden fencing to create a familiar, single-family home atmosphere.

ユニバーサル・トイレ入口付近。ミントグリーンのフレームが清々しく、壁には部分的に白い磁器タイルを用いた。

View near the entrance to the universal toilet room. The mint green frame is refreshing, and the walls are partially finished with white porcelain tiles.

男性用トイレの天井を見上げる。天井は吹き抜けで三角屋根の傾斜に沿って照明を取り付け、鉄骨の梁にシーリングファンを設置した。

Looking up at the ceiling of the men's toilet room. The ceiling is open, and lighting is mounted on the sloping triangular roof, with ceiling fans on the steel beams.

下左：歩道沿いの窓側に広くとった、男性用トイレの手洗い場。下右：ユニバーサル・トイレ内。内装は白が基調で清潔感がある。吹き抜けの天井により面積以上の広がりが感じられる。

Bottom left: The men's toilet room has a wide hand-washing counter by the window along the sidewalk. Bottom right: Interior view of the universal toilet room. The interior is basically white, giving a sense of cleanliness. The open ceiling creates a sense of spaciousness beyond the area.

右：神宮前1丁目交差点上に架かる、歩道橋からの外観。「原宿の片隅にひっそりと建つ」姿をイメージしてデザインされた。

Exterior view from the pedestrian bridge over the Jingumae 1-chome intersection. The design reflects the image of a building "standing quietly in a corner of Harajuku."

神宮前一丁目
Jingumae 1
消火栓

立面図。白い外壁にミントグリーンのフレーム、赤茶色の屋根といったカラーリングはすべて「ディペンデント・ハウス」から。

Elevation. The color palette of white exterior walls, a mint green frame, and a reddish brown roof are entirely drawn from the Dependents Housing.

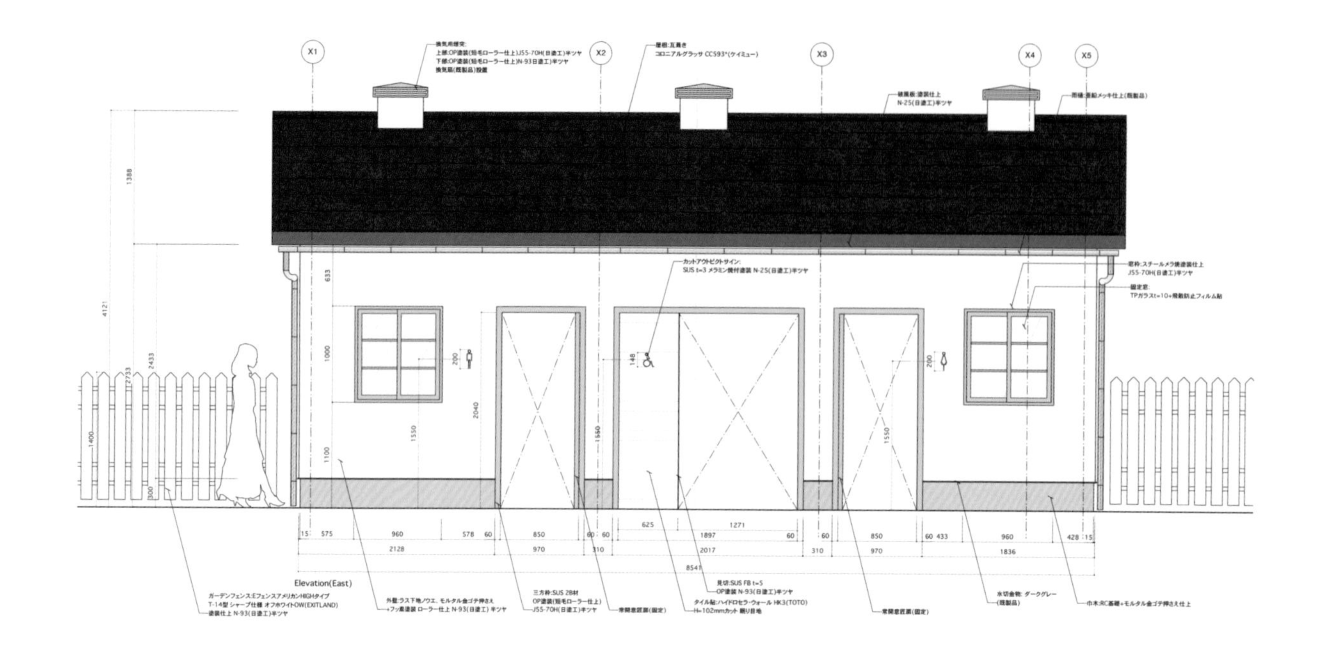

都市の隙間に佇む、懐かしくも新しい平屋の一軒家

A Nostalgic but New One-Story Detached House Nestled in a Corner of the City

神宮前1丁目の交差点は、東京で最も好きな交差点です。かつては毎日のように通ったショップやカレー店が近くにあったこともあり、ゆかりが深いこの場所の公共トイレに携われたことをうれしく思っています。

一軒家のような「神宮前公衆トイレ」のデザインの由来は、1946年に現在の代々木公園一帯に設けられた米軍駐留施設、ワシントンハイツの「ディペンデント・ハウス」です。ワシントンハイツは、原宿が現在のカルチャータウンに発展していくきっかけになった場所でもあり、また、「ディペンデント・ハウス」は、戦後の日本のライフスタイルの変化にも大きな影響を与えました。

その家も今では、ほとんど残っていません。私は、原宿に育てられて今があります。今回のトイレのコンセプトを「温故知新」とし、「ディペンデント・ハウス」のいわば“写し”に挑戦したのには、消えゆくデザインを大好きな街、原宿に残せたら、という思いもありました。世代によって懐かしくも感じ、また新しくも感じていただけると思います。

家型の親しみやすい建物であることが、公共トイレとして入りやすいものになるよう、内開きのドアが常に開いているように見せたり、素朴なガーデンフェンスのような柵にしたりと、細部の「ちょっとしたこと」にも気を配りました。

トイレとしての機能面は、デザイン前に講習を受け、さまざまな事を学んで取り組みました。使いやすさを第一に考え、構成はシンプルに、手洗い場のスペースを広くとり、衛生面についても非接触で使える水栓を選ぶなど、慎重に考えてデザインを進めました。公共トイレは助け合いの場所でもあると思っています。美しく、綺麗に、いつまでも。そんな気持ちで使ってもらえればと思います。

The Jingumae 1-chome intersection is my favorite intersection in Tokyo. Because there used to be stores and a curry shop that I went to daily nearby, I am happy to be involved in the public toilet facility construction in this place with which I have a deep connection.

Our design of the "Jingumae Public Toilet," which resembles a detached house, originated from the Dependents Housing at Washington Heights, a U.S. military stationing facility established in 1946 in the present-day Yoyogi Park area. Washington Heights was the catalyst for Harajuku's development into the cultural town it is today, and the Dependents Housing also greatly influenced the changing lifestyles of postwar Japan.

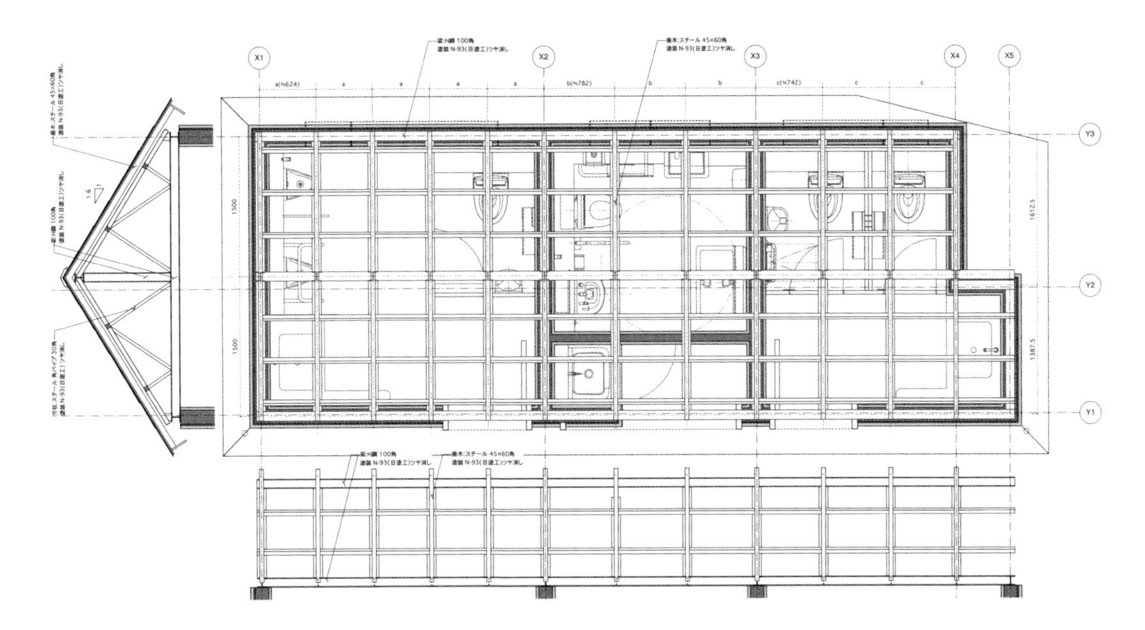

天井伏図。天井は小屋組が見える吹き抜けとした。建築面積は約22m²。個室1室と小便器2器の男性用トイレが約8.6m²と最も広い。

Ceiling Plan. The ceiling is a vaulted ceiling revealing the roof framing. The building area is approximately 22 m². The men's toilet room, with a cubicle and two urinals, is the largest room at about 8.6 m².

代々木公園の一角に残るディペンデント・ハウス。1964年の東京オリンピックではオランダ選手宿舎として利用され、オリンピック記念宿舎として保存されている。

Dependents Housing, part of which remains in Yoyogi Park, was used as a dormitory for the Dutch athletes at the 1964 Tokyo Olympics and is preserved as the Olympic Commemorative Athletes Housing.

Very few of those houses remain today. Harajuku is where I grew up and made me who I am today. I chose the concept of "Onko-chishin" ("Learning from the past to understand the new") for this facility, and challenged myself to make a "replica," so to speak, of the Dependents Housing, partly because I wanted to preserve this fading design in Harajuku, a town I love. This work may seem nostalgic to some or new to others, depending on their generation.

To make the familiar, house shaped building an accessible public toilet, we paid attention to small details, such as inward-opening doors that always appear open, and rustic, garden-like fencing.

We took a course to learn about various functional aspects of toilets before starting the design. Usability was our first priority, so we kept the configuration simple, made more space for hand wash basins, and chose non-contact faucets for hygienic use, all with careful consideration throughout the design process. We believe that public toilets are places where people help each other to keep them beautiful and clean forever. We hope that people will use them with that in mind.

Drawings and Specifications ▶P.254

NIGO®

ファッションデザイナー／クリエイティブディレクター。ストリートカルチャーのパイオニアとして知られ、現在は「ヒューマンメイド」のデザインをメインに多方面で活動。2021年9月、KENZOのアーティスティックディレクターに就任。

Fashion designer / creative director known as a pioneer of street culture. Currently active in many fields, primarily designing for Human Made. He was appointed Artistic Director of Kenzo in September 2021.

鍋島松濤公園トイレ

NABESHIMA SHOTO PARK PUBLIC TOILET

デザイン：隈研吾　渋谷区松濤2丁目10番7号

Design : Kengo Kuma　2-10-7 Shoto Shibuya-ku

2021年6月完成。鉄筋コンクリート造。閑静な住宅街にある公園の東、ベンチや遊具のある広場に面して建つ。分棟型で「公衆トイレの村」をイメージしてデザインされた。

Completed in June 2021. Reinforced concrete construction. It stands east of the park in a quiet residential area, facing a plaza with benches and playground equipment. The detached structures were designed as a "public toilet village."

左：階段を上がった正面に子ども用トイレがある。右上、右下：5つに分けたトイレを「森のコミチ」と名付けられた通路がつなぐ。通路にはウッドチップが敷き詰められている。

Left: An infant toilet is located at the top of the stepped pathway. Top right, bottom right: A path named "Mori no Komichi" (forest path) connects the five separate toilets. The path is paved with woodchips.

ユニバーサル・トイレ内。壁の装飾には、製材時に出る端材や古材を再利用した。
Interior view of the universal toilet. Scrap wood from lumbering and aged wood were reused for wall decoration.

身だしなみ配慮の男女共用トイレ内。鏡を大きく、手洗い器カウンターも広くとった。

Interior view of the unisex toilet designed for personal grooming. It features a large mirror and a wide hand-wash basin counter.

左上：ユニバーサル・トイレは公園の広場と段差のない場所に設置。左下：階段上、写真左手奥に小便器ブースがある。右：通路沿いに点在する照明が建物と足元を照らす。

Top left: The universal toilet is located flush with the park's plaza and steps. Bottom left: Urinal booths are located at the top of the stepped path, at the far-left side of the photo. Right: Lights scattered along the small path shine light on the buildings and one's feet.

隈研吾による「鍋島松濤公園トイレ」のスケッチ。森に溶け込む杉板の集落をイメージした。

A sketch of the Nabeshima Shoto Park Public Toilet by Kengo Kuma. He envisioned a cedar board village that blends into the forest.

小径の散策も楽しい、緑に溶け込む杉板の集落

A Cedar Board Village That Blends in with the Greenery, Offering a Pleasant Stroll along a Small Path

湧水池があり、森のような深い緑のある鍋島松濤公園にあって、その森に溶け込むような公共トイレをつくりたい、と思いました。そのため、大きな箱のような建物をつくるのではなく、5つの異なるトイレの小屋を点在させ、小屋と小屋の間を“森のコミチ”で結びました。

小屋の外壁は、ランダムな角度の杉板ルーバーで覆われています。この杉は年輪が緻密で高い強度をもつ、奈良県産の吉野杉。あえて木の耳を残した木材を使用したのは、森の中にいるような荒々しさを残したかったからです。

木を使うことは、都市の中に自然を取り戻すことだと考えています。今回のような公共の建物であっても、工夫次第でさまざまな木の使い方ができます。根底には、日本の木の文化を世界に発信したいという思いがあります。内装に用いたのは「さがみはら津久井産材」のサクラやケヤキ。机の天板を製材する際に出る小径木などを活用しています。

小屋と小屋を結ぶ小径をより自然に近い山道状にするため、公園にもともとあった勾配に沿った地形もデザインしました。これからの建築は、外との関係、人の感情との関係を含めたトータルな体験がより重要になってくると考えています。自然の中を歩きまわり、お気に入りの場所を見つけるように、自分が好きなトイレを選んで使ってもらえればうれしいです。

公共施設はこれまで数多く手掛けてきましたが、公園内の公共トイレを設計したのは、今回の「鍋島松濤公園トイレ」が初めてです。公共トイレはユニバーサルデザインという視点でみると、設計の密度を上げなくてはなりません。小さな建築ですが、大きな住宅や施設を設計するのと同じくらいの労力を費やしました。

車椅子を使用される方や介助が必要な方にもお使いいただけるユニバーサル・トイレをはじめ、幼児用便器のある子ども用トイレ、着替え台のある身だしなみ配慮トイレなど、5つのトイレはそれぞれに、利用者のさまざまなニーズに沿う作り方をしています。「THE TOKYO TOILET」は多様性を受け入れる社会の実現を目的にしていますが、「鍋島松濤公園トイレ」の設計テーマとして大切にしたことのひとつも、まさに「多様性」です。機能も内装も異なる5つのトイレは、多様性の時代を象徴するものでもあります。

Nabeshima Shoto Park has a spring-fed pond and deep forest-like greenery. Our idea was to create a public toilet blending in with the forest. Instead of building a large box-like structure, we built five different toilet huts scattered throughout the park, with a "Mori no Komichi" (forest trail) connecting them.

The exterior walls of the huts are covered with randomly angled cedar louvers. They are made of Yoshino cedar from Nara Prefecture, known for its dense growth rings and high strength. We used live edge boards to create a rough, natural

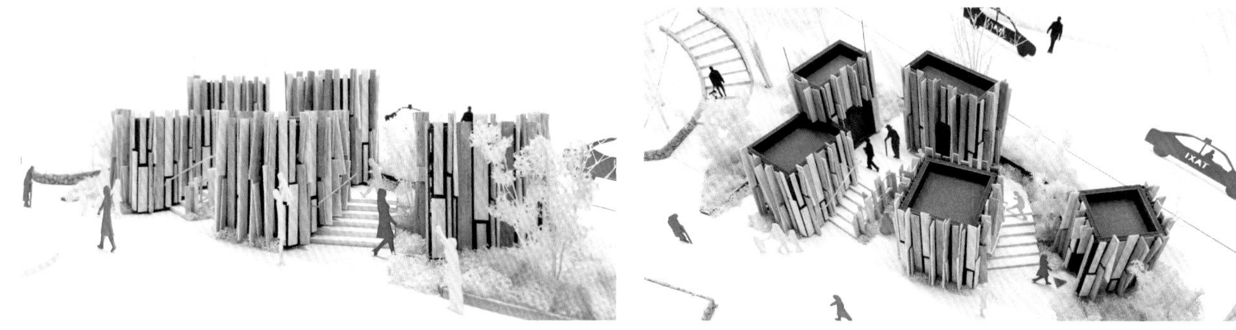

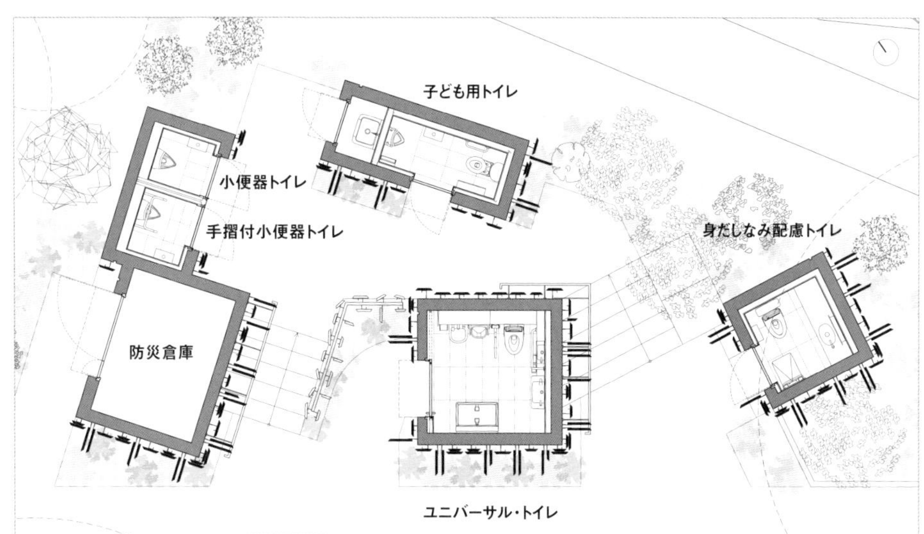

機能を分散した5つの棟の配置模型と平面図。敷地面積は約394m²と公園内の公共トイレとしては比較的広いが、建物のボリュームを極力抑え、建築面積は5棟あわせて約21m²に。公園の樹木との一体感を意図し、外壁には約240枚の、長短さまざまな木材を組み合わせた。

Layout model and floor plan of the five buildings with separate functions. While the site area is about 394m² relatively large for a public toilet in a park, the building area was minimized to a total of around 21m² by reducing the building volume as much as possible. The exterior walls are made up of roughly 240 wooden boards of varying length and width, designed to blend in with the trees in the park.

forest-like appearance.

We believe that using trees brings nature back into the city. Even in public buildings like this one, trees can be used in a variety of ways, with some ingenuity. One of our underlying aspirations is to share Japan's wood culture with the world. The interior is made of cherry and zelkova wood from "Sagamihara Tsukui," utilizing small-diameter wood generated from sawing desk tops.

To make the path connecting the huts more natural and mountainous, we also designed a topography conforming to the original slope of the park. We believe that architecture in the future will increasingly involve a total experience including the relationship with the outside and people's emotions. We invite you to choose and use the restroom of your choice, just like wandering through nature and finding your favorite place.

"Nabeshima Shoto Park Public Toilet" is the first public toilet we have designed for a park, among many public facilities we have worked on in the past. From a universal design perspective, public toilets require increased design density. While it is a small building, we put the same amount of effort into it as we would have in designing larger residences or facilities.

Each of the five toilets, including a universal toilet suitable for wheelchair users and those who require assistance, a children's toilet with an infant urinal, and a grooming toilet with a changing table, is designed to meet users' various needs. THE TOKYO TOILET project aims to realize a society that embraces diversity, and one of the important design themes for the "Nabeshima Shoto Park Public Toilet" is precisely "diversity". The five toilets, each with a different function and interior, symbolize the era of diversity.

Drawings and Specifications ▶P.256

隈研吾　Kengo Kuma

建築家。1954年生まれ。90年隈研吾建築都市設計事務所設立。木をはじめ自然素材を大胆に用いた作品で注目され、世界各地で数多くのプロジェクトが進行中。2021年に給付型奨学金を提供する、公益財団法人隈研吾建築奨学財団設立。

Architect, born in 1954. Established Kengo Kuma & Associates in 1990. Renowned for his bold use of wood and other natural materials, Kuma has numerous projects underway around the world. In 2021, he established the Kengo Kuma Foundation, a public interest incorporated foundation offering benefit scholarships.

恵比寿駅前交番周辺案内図

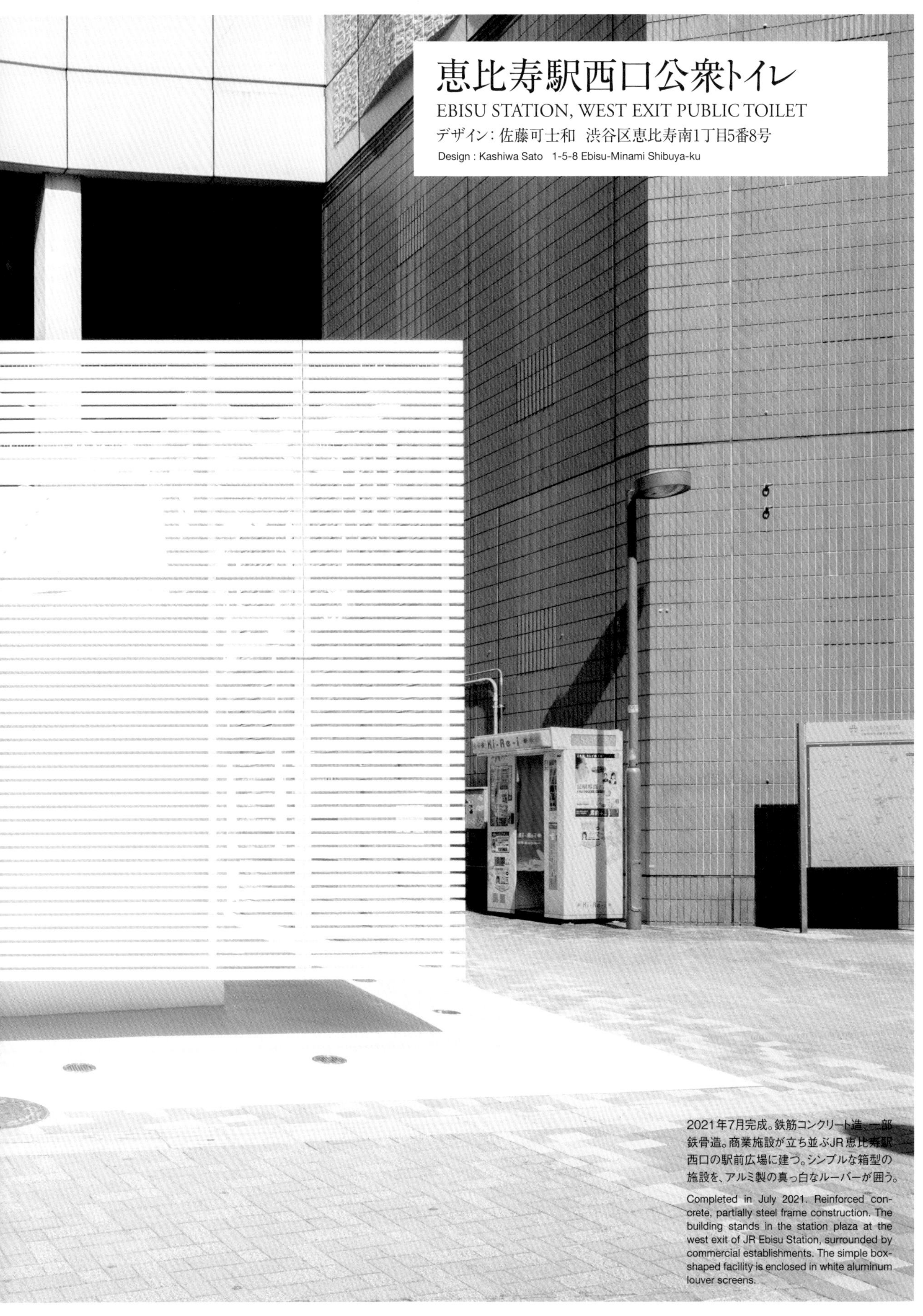

恵比寿駅西口公衆トイレ

EBISU STATION, WEST EXIT PUBLIC TOILET

デザイン：佐藤可士和　渋谷区恵比寿南1丁目5番8号

Design : Kashiwa Sato　1-5-8 Ebisu-Minami Shibuya-ku

2021年7月完成。鉄筋コンクリート造、一部鉄骨造。商業施設が立ち並ぶJR恵比寿駅西口の駅前広場に建つ。シンプルな箱型の施設を、アルミ製の真っ白なルーバーが囲う。

Completed in July 2021. Reinforced concrete, partially steel frame construction. The building stands in the station plaza at the west exit of JR Ebisu Station, surrounded by commercial establishments. The simple box-shaped facility is enclosed in white aluminum louver screens.

左：通路入口からの見通し。右：コの字に囲ったルーバーの端、左右2カ所に、個室へと至る通路の出入口がある。ユニバーサル・トイレへは、通路を経ずにアクセスできる。

Left: View from the passageway entrance. Right: At both ends of the U-shaped louver screens, there are two entrances to the passageway leading to the toilet rooms on either side. The universal toilet rooms can be accessed without entering the passageway.

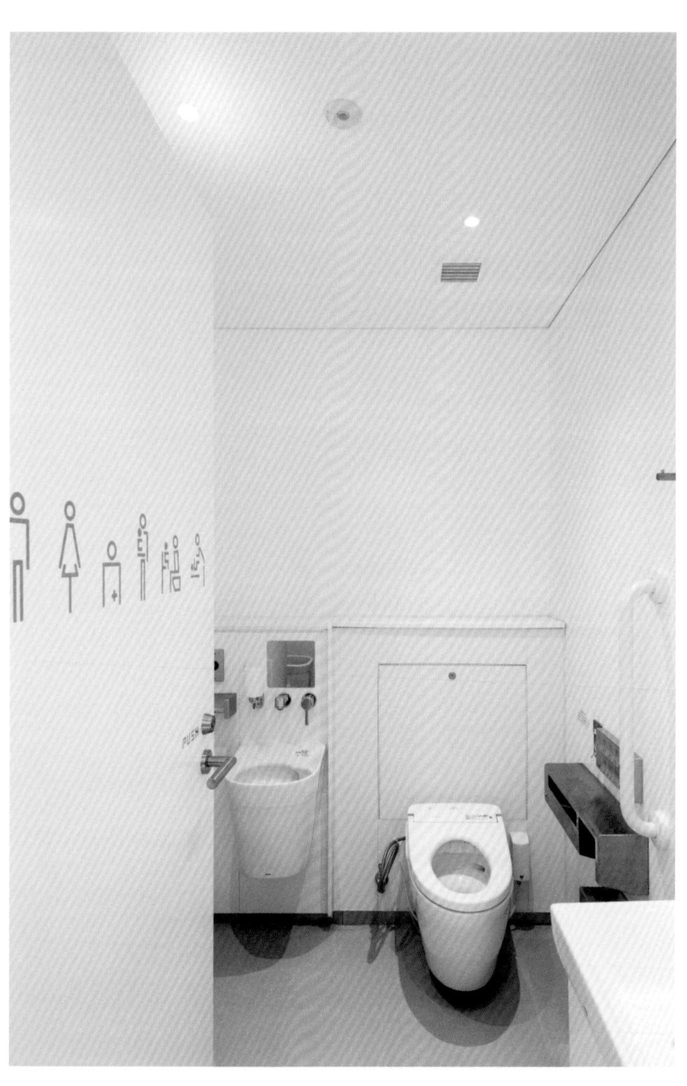
PUSH

左、右：5室ある個室はすべて男女共用で、ユニバーサル・トイレのほかに、ベビーシートとベビーチェアなどを備えた乳幼児連れ配慮トイレ、オストメイト対応トイレも設けている。

Left, right: All five toilet rooms are gender-free, featuring a universal toilet, an infant-friendly toilet equipped with a baby changing station and baby chair, and an ostomate-accessible toilet.

夕景。床に埋め込んだアッパーライトが建物全体を照らす。ルーバーの足元は50cmほど開け、施設内の通路を歩く人の様子が歩道からもわかるようにした。建築面積は約19m²。

Evening view. Uplights embedded in the floor illuminate the entire building. The louver screens are raised 50 cm above the ground so that people walking along the passageway in the toilet facility are visible from the sidewalk. The building area is about 19m².

JR
JR東日本
恵比寿駅
Ebisu Station

JRと東京メトロの乗り換え通路付近から見た外観夕景。照明の演出で軽やかさが際立つ。

Evening view of the exterior from the transit passage between the JR and Tokyo metro lines. The lighting enhances the lightweight appearance.

「あたりまえの配慮」に向き合い、白く、軽やかに、目立ち過ぎず

A White, Light, and Unobtrusive Facility Addressing "Common Sense Considerations"

「THE TOKYO TOILET」としてデザインされた公共トイレは、公園内に建つもの、大通りの近くに建つものなど、それぞれ異なる立地環境にあります。場所によって求められることは違います。ゆえにデザインにも違いがあり、違うからこそ、面白い。

「恵比寿駅西口公衆トイレ」周辺は人通りが多く、敷地のすぐ近くには、待ち合わせによく使われる「えびす像」もあります。いわば街の顔とも言える場所で、恵比寿駅西口を最寄り駅とする人にとっては、毎日目にする建物にもなります。駅前に建つ公共トイレはどうあるべきか。まずはこの場所特有の立地条件を整理し、トイレとして目立つことよりも、"目立ち過ぎない"ことを大切にしたい、と考えました。

いい意味で気にならない、一歩引いた佇まい。白を選んだのは、清潔さを伝えると同時に、駅前の雰囲気が少し明るく、清々しいものになれば、と思ってのことです。ルーバーはプライバシーの確保と外部からのほどよい視認性という2つの「安心」を両立させるためのもので、風通しもよく、全体の印象を軽やかなものにしてくれています。

また、建物だけではなく、敷地部分も白の塗り床仕上げにすることで、昼は床からの反射光でフワリとした存在感を。夜は下からのライティングで建物全体が行燈のような光をまとい、周囲を照らす計画としました。どちらも魅力的な「柔らかな白」が実現できたと思っています。

計画全体を通して心がけたのは、清潔で使いやすいという公共トイレとしての"あたりまえの配慮"に一つひとつ向き合い、新しさと違和感のなさのちょうどいい"さじ加減"を探ること。「恵比寿駅西口公衆トイレ」が利用者に違和感なく、新鮮な気持ちで受け止めてもらえることを願っています。

The public toilets designed as part of "THE TOKYO TOILET" project are located in various environments, with some in parks and others near main streets. The requirements differ depending on the location. Therefore, the designs vary considerably, and this is what makes them so interesting.

The area around the Ebisu Station West Exit Public Toilet is bust with passersby, and the statue of Ebisu, often used as a meeting place, stands near the site. It is the public face of the district, and those commuting through the west exit of Ebisu Station will see the building daily. What should a public toilet facility standing in front of a station look like? First, we examined the unique conditions of this location and decided that the facility should appear "unobtrusive" rather than prominent.

鉄筋コンクリート造の施設本体を囲うルーバーは、原寸模型で高さなどを検討した。

The height of the louvers around the main concrete structure was determined using a full-scale model.

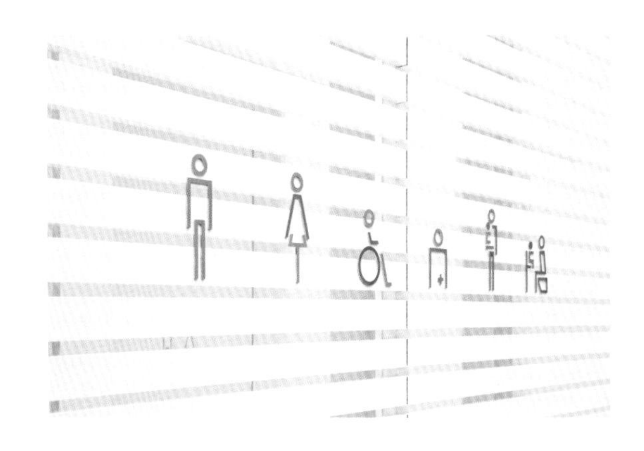

「THE TOKYO TOILET」のすべてのピクトグラムは佐藤可士和によるデザイン。各トイレにそれぞれのクリエイター名を記したプレートも設置されている。

Kashiwa Sato designed all of the pictograms in THE TOKYO TOILET facilities. Each toilet facility has a plaque with the creator's name on it.

It should have a discreet, unassuming appearance, not attracting attention in a positive sense. We chose the color white in the hope that it would convey cleanliness and, at the same time, brighten and refresh the atmosphere in front of the station a little. The louvers are designed to offer two types of reassurances: privacy and moderate visibility from the outside, providing good ventilation and a lighter overall impression.

In addition to the building, we painted the ground of the site area white to achieve an airy impression by reflecting light from the floor during the daytime. The entire building is lit from below at night, illuminating the surrounding area with a lantern-like glow. We believe we have succeeded in creating an appealing "soft whiteness" at night and day.

Our goal throughout the project was to find the right balance between novelty and comfort by addressing cleanliness and usability, the "common sense considerations" of public toilets, one by one. We hope that the "Ebisu Station West Exit Public Toilet" will be accepted by users with a fresh feeling and no sense of discomfort.

MAP 15

Drawings and Specifications ▶P.258

佐藤可士和 Kashiwa Sato

クリエイティブディレクター。1965年生まれ。博報堂を経て2000年 SAMURAI設立。グラフィックデザインのみならず、ブランド戦略のトータルプロデューサーとしてコンセプトの構築からコミュニケーション計画の設計、ビジュアル開発まで行う。

Creative director, born in 1965. After working at Hakuhodo, Kashiwa Sato established SAMURAI Inc. in 2000. In addition to graphic design, he also works as a total producer of brand strategy, from concept development to communication planning and visual development.

代々木八幡公衆トイレ

YOYOGI-HACHIMAN PUBLIC TOILET

デザイン：伊東豊雄　渋谷区代々木5丁目1番2号

Design : Toyo Ito　5-1-2 Yoyogi Shibuya-ku

2021年7月完成。鉄筋コンクリート造。左から女性用、ユニバーサル、男性用トイレがそれぞれ独立して建つ。周囲を回遊できる分棟計画で死角をなくし安全性を高めた。

Completed in July 2021. Reinforced concrete construction. The women's, universal, and men's toilet rooms, from left to right, stand apart from each other. The dispersed layout of the detached structures eliminated dead ends and enhanced safety.

代々木八幡宮
車路
住居表示街区案内

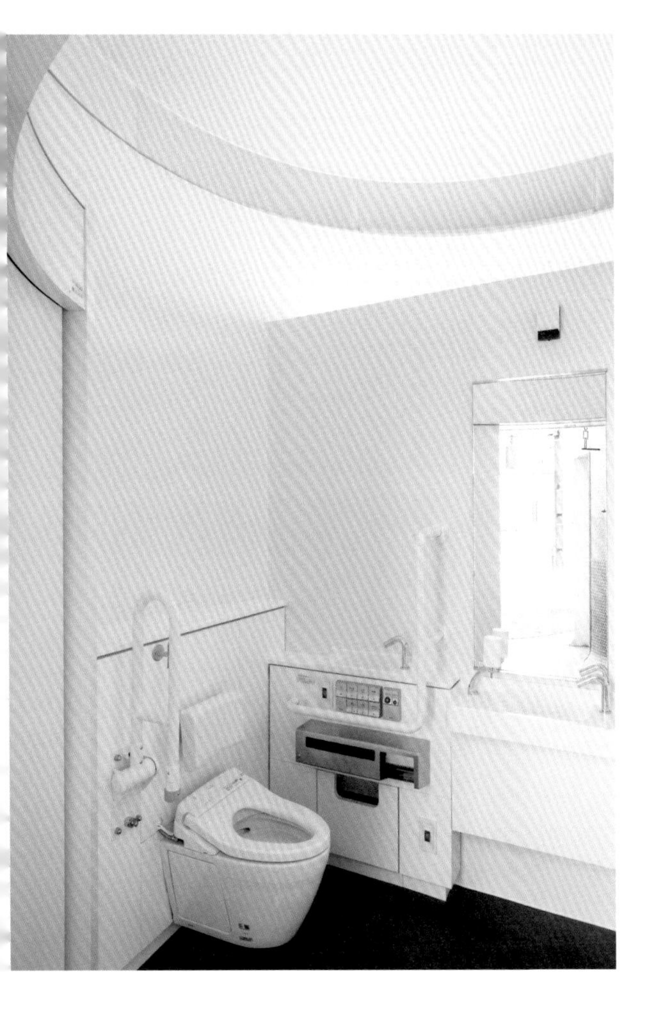

左：男性用トイレの個室背面を小便器ゾーンとし、壁を立てて水栓をつけた。上左：ユニバーサル・トイレ内。天井下のガラス窓から自然光が差し込む。上右：個室2室が並ぶ双子のキノコのような女性用トイレ外観。

Left: The back of the men's toilet room is a urinal zone with a detached wall equipped with a hand wash basin. Top left: Interior view of the universal toilet room. Natural light pours in through a glass window under the ceiling. Top right: Exterior view of the women's toilet room, with two cubicles arranged side by side like twin mushrooms.

上：建物本体とガラス窓の上にステンレスの屋根が浮かぶようにかかる。下左、中：外壁のモザイクタイルは、セルフクリーニング機能を施した特注品。下右：個室内から。自然光が差し込むガラス窓を見上げる。

Top: Stainless steel roofs hang as if floating above the structures and glass windows. Bottom left, middle: The mosaic tiles on the exterior walls are custom-made with self-cleaning features. Bottom right: View from inside a cubicle. Looking up at the glass window letting in natural light.

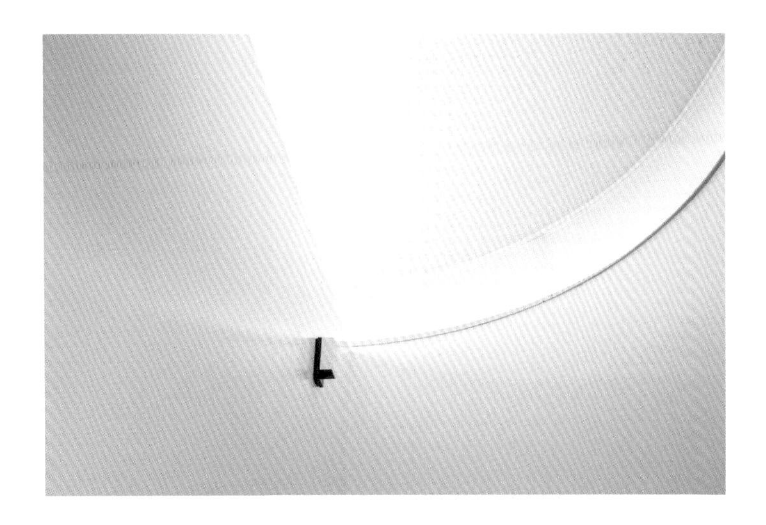

左：個室扉の外部照明のほか、夜は各棟のガラス窓からの光が屋根裏の軒に反射し、敷地内を明るく照らす。右：ユニバーサル・トイレ出入口付近から、夜の山手通りを見通す。

Left: Exterior lighting on the toilet room doors and light from the glass windows of each toilet room reflect off the attic eaves and illuminates the site at night. Right: Overlooking Yamate Street at night from near the entrance to the universal toilet room.

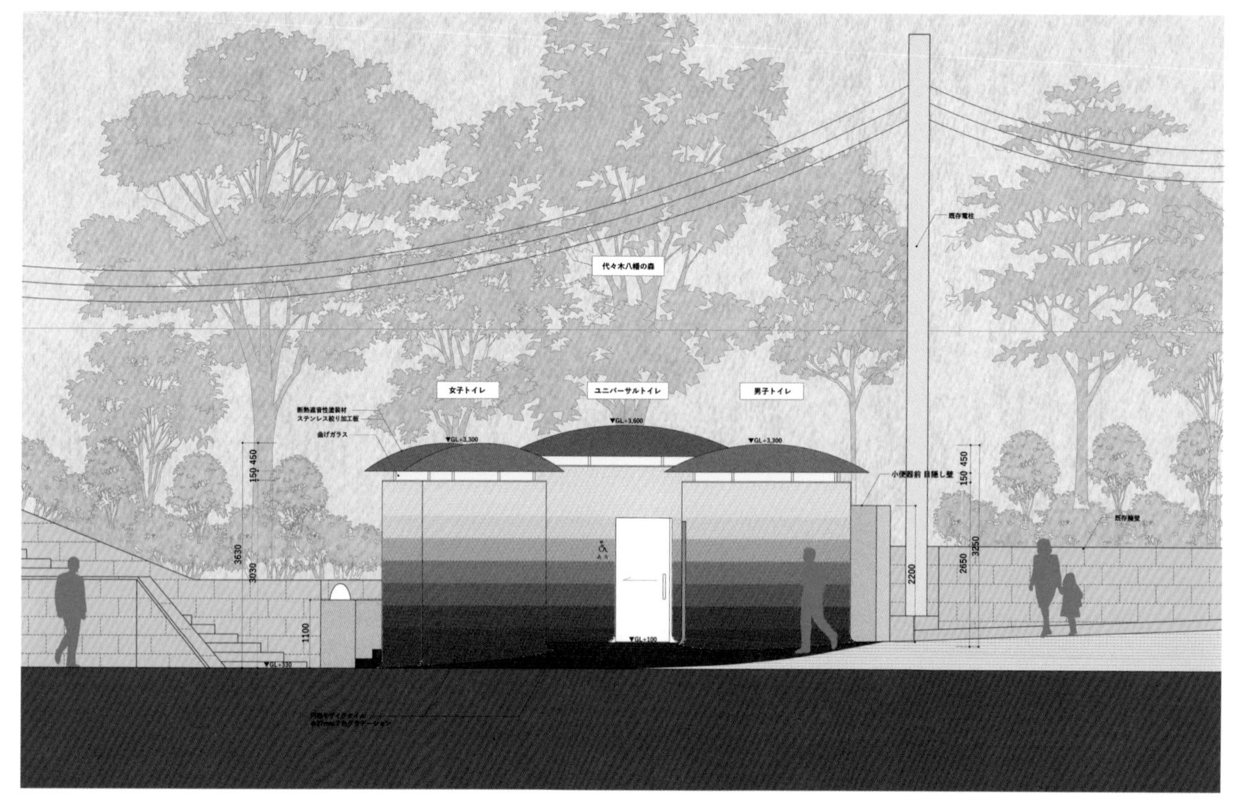

立面図。代々木八幡宮の参道にあたる階段の上り口横にある。建築面積は約17.5m²。最も広い中央のユニバーサル・トイレの床面積は約7.8m²。

Elevation. The structures are located next to the stairs of the approach to the Yoyogi Hachiman Shrine. The building area is about 17.5m². The largest structure, the universal toilet room in the center, has a floor space of about 7.8m².

森のキノコを思わせる、やさしさと入りやすさのデザイン

Design Offering Kindness and Accessibility, Reminiscent of Mushrooms in the Forest

敷地は、創建から800年以上の歴史をもつ代々木八幡宮の森を背にした、三角の変形敷地。交通量や人通りの多い環状6号線（山手通り）に面していながら、建替え前のトイレは暗く、入りにくい印象がありました。そのため、まずは明るく、入りやすく、特に女性が夜間でも安心して利用できる公共トイレにしたい、という思いがありました。

個室型のトイレを3つに分散させることで回遊性を生み出し、行き止まりがなく視線が抜けることで、防犯性を高めています。また、従来はユニバーサル・トイレに集約されていた子ども連れや高齢者のための機能を、男女の個室にもそれぞれ備え、3棟ともに多様なニーズに応えられる仕様になっています。多くの公共トイレが清潔に保つことが難しい状況を鑑み、換気をよくしたり落書きを防いだりすることにも、とりわけ配慮して設計をしました。

円柱状の室内や、浮かぶようにかけた丸みのある屋根など、建物の形状は「キノコ」を連想させる表現を採りました。でも、最初からキノコにしたかったわけではありません。地面から生えてきたような建物にしたい、とは思っていました。ダークブラウンからオフホワイトへのグラデーションで並ぶ外壁のモザイクタイルは、大地のエネルギーが建物に伝わり、上に行くにしたがって空に溶けていくようなイメージを表現しています。

公共建築をつくるときには、できるだけ自然の中にそのままいるような建物にしたいと考えています。周りの環境と切り離された場所ではなく、利用者が緊張せずに心地よく過ごせる空間にしたい。キノコのようなこの「代々木八幡公衆トイレ」が、森と調和し、子どもたちをはじめ、地域の方々によろこんで使ってもらえたらうれしいです。

The site is an irregularly shaped triangular lot set against the forest of Yoyogi Hachiman Shrine, with an 800-plus-year history since its construction. The toilet before reconstruction was dark and appeared uninviting, despite its location on the busy Loop Route 6 (Yamate Dori). For this reason, our first priority was to provide a public toilet that is well-lit, easy to enter, and especially safe for women to use at night.

Dispersing three stand-alone toilet rooms enhances security by facilitating circulation and providing a line of sight without dead ends. All three rooms, including the men's and women's, are equipped with functions to meet diverse needs of those accompanying children and the elderly, which were previously concentrated in the universal toilet. Considering the difficulties

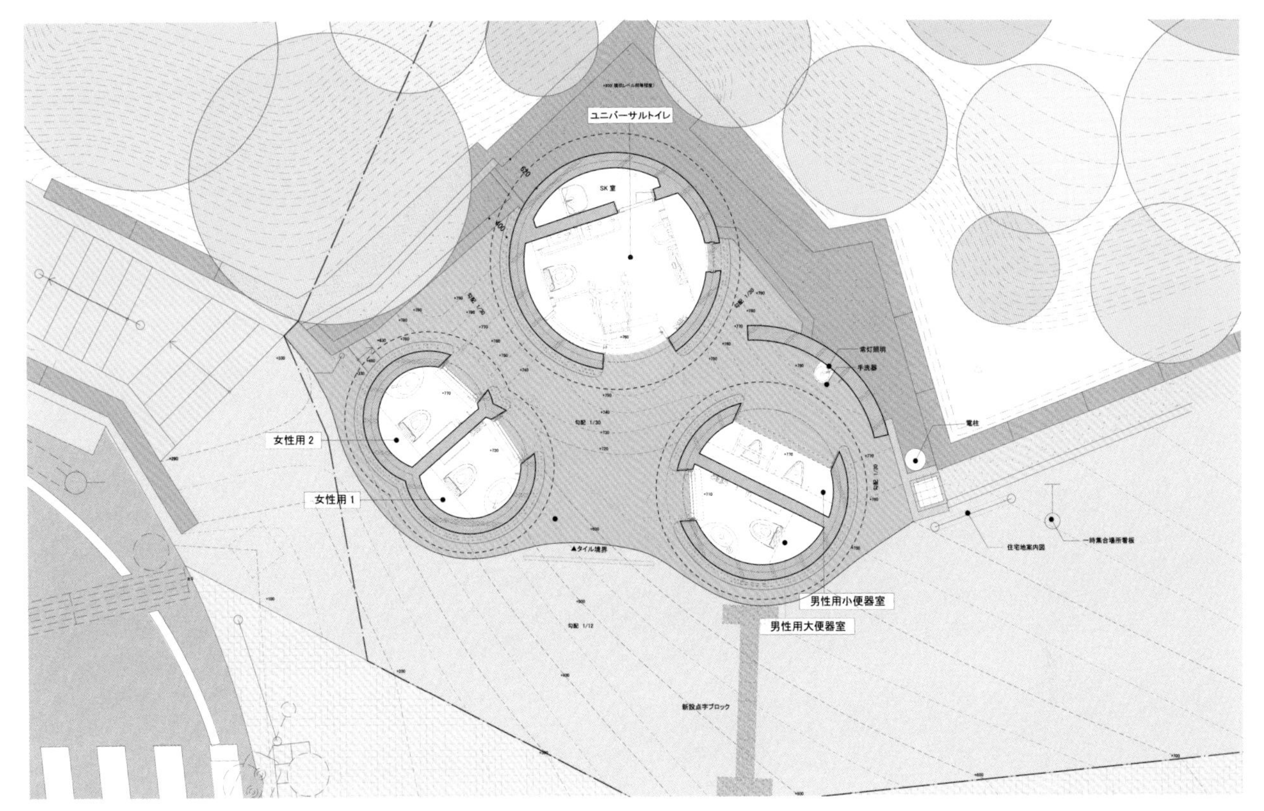

行き止まりがなく視線が抜ける平面計画。男性用、女性用トイレともに手洗い場も室内に備え、双方にベビーチェアを設置した。

Floor plan without dead ends and with clear visibility. Both the men's and women's toilet room are equipped with hand wash basins inside, along with baby chairs.

many public toilet facilities have in maintaining cleanliness, our design also paid particular attention to facilitating good ventilation and graffiti prevention. The shape of the rooms, with cylindrical interiors and rounded roofs that seem to float, evokes "mushrooms." However, making the rooms look like mushrooms was not our initial idea. What we wanted was to make the rooms look as if they had grown out of the ground. The mosaic tiles on the exterior walls, arranged in gradations from dark brown to off-white, express the image of the earth's energy transferring to the toilet rooms and dissolving into the sky as it moves upward.

When designing a public facility, we strive to give it a feeling of being directly in nature as much as possible. Our goal is to create a space that is not isolated from the surrounding environment, but a space where users can feel comfortable without any tension. We hope the mushroom-like "Yoyogi-Hachiman Public Toilet" will blend in with the forest, and children and other community residents will enjoy using it.

MAP 06
Drawings and Specifications ▶P.260

伊東豊雄 Toyo Ito

建築家。1941年生まれ。菊竹清訓建築設計事務所を経て71年、現・伊東豊雄建築設計事務所設立。国内の代表作に「せんだいメディアテーク」「みんなの森 ぎふメディアコスモス」など。2011年にこれからのまちや建築を考える場として、私塾「伊東建築塾」設立。

Architect, born in 1941. After working at Kiyonori Kikutake & Associates, Toyo Ito established the present Toyo Ito & Associates in 1971. Representative works in Japan include Sendai Mediatheque and Minna no Mori Gifu Media Cosmos. In 2011, he established a private school "Ito Kenchiku Juku" as a place to think about the future of cities and architecture.

2021年8月完成。鉄筋コンクリート造。敷地北側の水道通りから見ると、真っ白な球体が浮かんでいるように見える。建替えと同時に施設周囲も白く舗装した。

Completed in August 2021. Reinforced concrete construction. When viewed from Suido Dori on the north side of the site, the building resembles a pure white sphere floating in the air. The floor around the facility was repaved in white as part of the reconstruction.

七号通り公園トイレ

NANAGO DORI PARK PUBLIC TOILET

デザイン: 佐藤カズー/Disruption Lab Team　渋谷区幡ヶ谷2丁目53番5号

Design : Kazoo Sato / Disruption Lab Team　2-53-5 Hatagaya Shibuya-ku

左上、左下：球体断面の左側がユニバーサル・トイレ。入口横の壁のモニターに表示されたQRコードをスマートフォンで読み込むと、施設機器を音声操作できる。右：男性用トイレ入口。衝立の奥に小便器2器を備える。

Top left, bottom left: Cross-section of the sphere shows the universal toilet room on the left side. By reading the QR code displayed on the wall monitor next to the entrance with a smartphone, users can operate the facility's equipment by voice. Right: View of the entrance to the men's toilet. Two urinals are located behind the partition.

Say "Hi Toilet",
then after the chime,
state your command.
Flush Toilet
Start Washlet
Stop Washlet
More Pressure
Less Pressure
Start Bidet
Stop Bidet
More Pressure
Less Pressure
Move Forward
Move Backward
Start Dryer
Stop Dryer
Play Music
Play Therapeutic Music
Stop Music

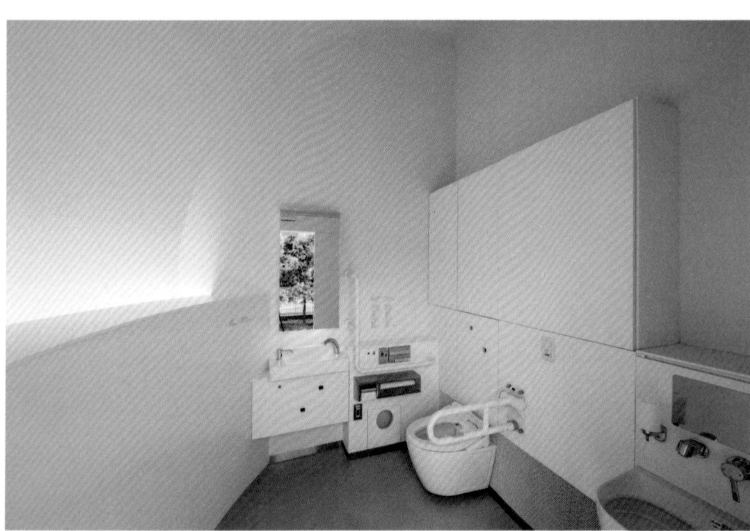

左上、左下：ユニバーサル・トイレ内。リモコンスイッチ横にマイクが仕込まれており、すぐ上の壁に音声認識可能な言葉が記されている。右：男性用トイレ内。厚みのある腰壁は24時間換気システムの給気口でもある。

Top left, bottom left: Interior view of the universal toilet room. A microphone is installed next to the remote-control switch, and words that can be recognized by the voice recognition device are written on the wall above it. Right: Interior view of the men's toilet room. The thick spandrel wall also serves as the air intake for the 24-hour ventilation system.

左：夜は照明の反射で浮遊感が際立つ。右上：スマートフォンでの読み込みが完了し、音声操作が可能となると、扉のフレームのライトカラーが青に変わる。右下：施設下には球形の輪郭に沿って白い砂利を敷き詰めた。

Left: At night, the reflection of the lighting enhances the floating effect. Top right: When a smartphone loads the QR code and activates voice control, the color of the lighting on the door frame changes to blue. Bottom right: White gravel is spread along the spherical outline around the bottom of the facility.

月　日

部

発行所

TOTO出版

著者

[文]岡野民

[写真]永禮賢

価3,300円

体3,000円+税10%)

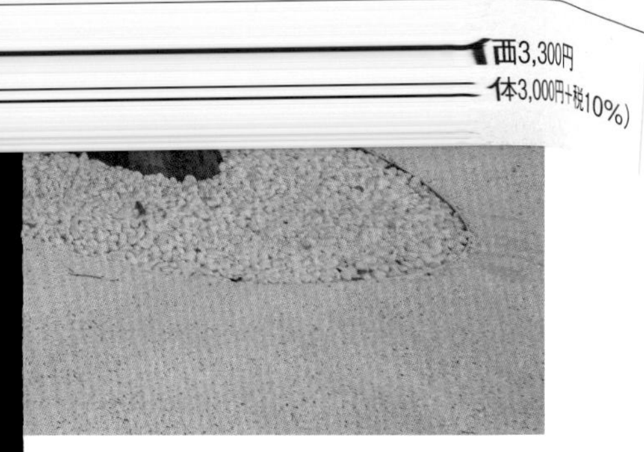

「Hi Toilet」と呼びかけると音声認識システムが始動。手を使わずに利用できるボイスコマンド式トイレのコンセプトを記したスケッチ。

Calling out "Hi Toilet" activates the voice command system. Sketch describing the concept of a hands-free voice-command toilet.

動作を声で指示する、最先端の〝手を使わない〟トイレ

State-of-the-Art "Hands-Free" Toilets with Voice-Command Movement Instructions

目指したのは、世界一清潔な公共トイレ。その目標にどう近づけるか、考えに考えた結果、ボイスコマンド式の"手を使わない"トイレに辿り着きました。

きっかけは、欧米のとあるトイレに関しての調査結果を見たことでした。それによると、トイレ利用者の約60%が洗浄レバーを足で踏んで水を流し、約50%がトイレットペーパーを用いてドアを開け、約40%がお尻でドアを閉め、約30%が肘を使い手の接触を避けるといいます。この"どこにも触れたくない"という心理を、新しいUX(ユーザー・エクスペリエンス)としてデザインしよう、と考えました。

そこからリサーチ&プランニングに3年をかけ、ボイスコマンドのテクニカルディレクターや建築の専門家の協力を得て実現したのが、この「七号通り公園トイレ」です。名前は「Hi Toilet」。たとえば「扉を開けて」「トイレの水を流して」など、トイレを利用する際に必要となる動作を声で指示すれば、非接触で用が済ませます。

言語は日・英の2カ国語対応。リラックスして利用してもらえるよう、あらかじめプログラムされた音楽もリクエストできます。もちろん音声だけではなく、これまで通り手を使ってのドアの開閉や便器の操作も可能です。

個室は、敷地周辺のニーズと建替え前のトイレの利用状況から、ユニバーサル・トイレだけでなく男性用トイレの2つとしました。

最大約4mの天井高をもつ真っ白な球形の建物は、一粒の水が落ちてきたような、瑞々しさをイメージしています。コンクリートの構造体で球体を作るのはとても難しいのですが、球体は、空気の流れを制御し、においが滞留しないための形でもあり、自然給気と機械排気を組み合わせた24時間換気システムを導入しています。

Our goal was to create the world's cleanest public toilet. Through much thought and consideration, we came up with voice-command hands-free toilets.

Our initial inspiration came from a survey of toilet facilities in the United States and Europe. According to the survey, some 60% of toilet users use their feet to flush the toilet, 50% open the door using toilet paper, 40% close the door with their buttocks, and 30% avoid hand contact by using their elbows. We decided to design a new user experience based on this "I don't want to touch anywhere" mentality.

After three years of research and planning from that point, and with the help of the technical director of voice command

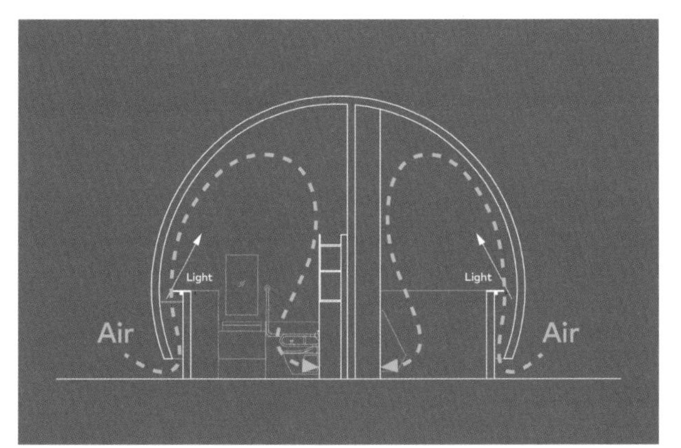

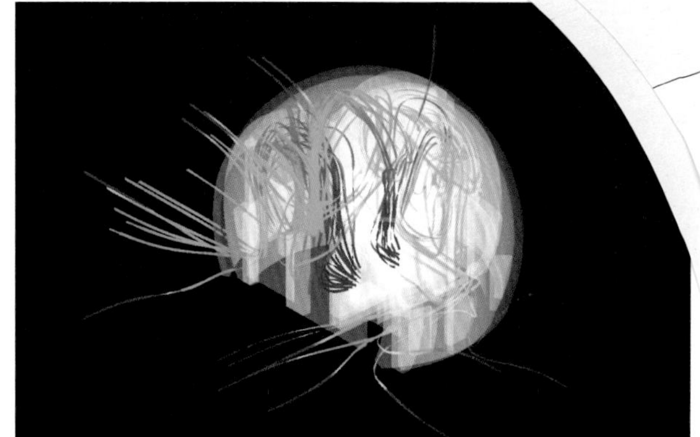

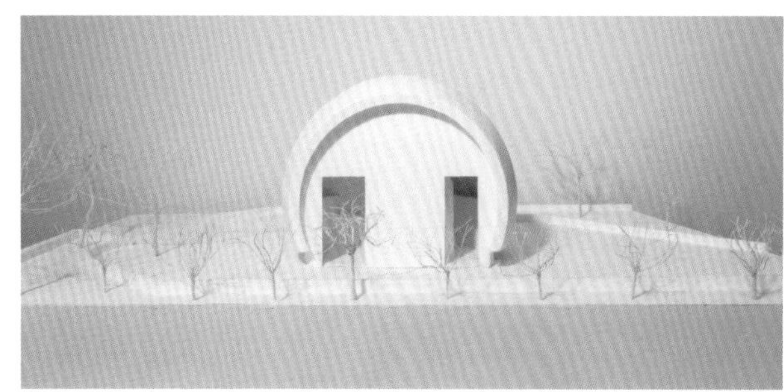

24時間換気システムによる空気循環のイメージ図と全体模型。空気の取り込みは外壁下部から。取り込んだ空気は球体のドーム天井を伝い、衛生設備機器の背後に収めたファンで排出している。天井高は最大で約4m。

Diagram and overall model of air circulation through a 24-hour ventilation system. Air is drawn in from the lower part of the exterior wall and then discharged through the spherical dome ceiling using a fan mounted behind the sanitary equipment. The maximum ceiling height is roughly 4 m.

and architectural experts, the "Nanago Dori Park Public Toilet" was realized. The name is "Hi Toilet." For example, by giving voice commands such as "open the door" or "flush the toilet," users can easily use the toilet without any contact.

Two languages, Japanese and English, are available for this function, and a pre-programmed music can be requested for a relaxing experience. Of course, you can use your hands to open and close the door and operate the toilets in the conventional way.

We decided it should have two toilet rooms, a universal toilet and a men's toilet rooms, based on the needs around the site and the status of toilet use prior to reconstruction.

The pure white spherical building, with a ceiling height of up to 4m, is designed to evoke freshness like a single drop of water falling on the ground. A sphere, a difficult shape to build with concrete, is also a shape that is able to controls air flow and prevent odors from lingering. The building features a 24-hour ventilation system combining natural air supply and mechanical air exhaust.

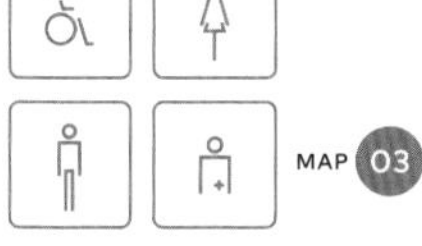

MAP 03

Drawings and Specifications ▶P.262

佐藤カズー Kazoo Sato

クリエイティブディレクター。1973年生まれ。ソニー・ミュージックエンタテインメント、レオ・バーネットを経て、2010年からTBWA\HAKUHODO。広告の分野で活躍するほか、ミュージックビデオやコンサートのアートディレクションなども行う。

Creative director, born in 1973. After working for Sony Music Entertainment and Leo Burnett, he joined TBWA\HAKUHODO in 2010. Along with his advertising work, he also engages in art direction for music videos and concerts.

広尾東公園トイレ

HIROO HIGASHI PARK PUBLIC TOILET

デザイン：後智仁　渋谷区広尾4丁目2番27号

Design : Tomohito Ushiro　4-2-27 Hiroo Shibuya-ku

2022年7月完成。鉄筋コンクリート造。建物入口の反対側に、6m近い幅の間接照明パネルを設置した。パネルには79億通りのライティングパターンが映し出される。

Completed in July 2022. Reinforced concrete construction. An indirect lighting panel nearly 6 meters wide was installed at the rear of the building opposite the toilet entrances. The panel projects 7.9 billion different lighting patterns.

左上、右：南面の外壁はコンクリート打ち放し。2つのユニバーサル・トイレとした。左下：正方形に近いシンプルな室内。ゆとりのある広さで車椅子の取り回しもしやすい。

Top left, right: The exterior wall on the south side is exposed concrete. The facility consists of two universal toilet rooms. Bottom left: A simple, nearly square interior. The room is spacious and easy to maneuver for a person in a wheelchair.

左：植栽にも力を入れた。植栽デザインは「叢-Qusamura」の小田康平。右：敷地北側の歩道からの外観。日中のパネルのパターンは淡いグリーン。公園の木々とも調和する。

Left: Kohei Oda of Qusamura designed the planting with careful consideration. Right: View from the sidewalk on the north side of the site. During the day, the panel projects patterns in pale green. It blends well with the trees in the park.

左：ガラススクリーンと間接照明パネルの間。夜は白色で蛍や月明かりをイメージした。右上：美しい木漏れ日にも似た淡いグリーンのパターン。右下：南側の夜の外観。

Left: View between the glass screen and the indirect lighting panel. The white color at night evokes fireflies and moonlight. Top right: The pale green pattern resembles beautiful sunlight filtering through the trees. Bottom right: Exterior view of the south side of the facility at night.

北面のイメージパース。79億通りのパターンが一巡するには、計算上6000年以上かかるため、同じパターンを見ることはない。

Perspective view of the north elevation. Since 7.9 billion patterns will take over six thousand years to play out in a complete cycle, the same pattern will never appear twice.

79億通りのパターンを映す、トイレとアートのハイブリッド

A Hybrid of Toilet and Art Projecting 7.9 Billion Different Patterns

敷地は広尾ガーデンヒルズへと続く並木道の途中で、背後には聖心女子大学の校舎があります。そのような立地環境に建つ「広尾東公園トイレ」のデザインには、長く暮らしている住民の方々をはじめ、利用者の生活の一部になることへの責任を常に感じながら取り組んできました。

最初に考えたことは、どうしたらトイレがもっと好かれるか。大切にされるか。きれいなトイレだから好き、なのではなく、トイレ自体をもっと積極的に好きになる仕組みや「街との新しい関係性」がつくれないものか、ということでした。

パブリックアートは多くの場合、皆さんに愛され、大切にされていますよね。トイレを汚す人はいても、パブリックアートをないがしろにする人は少ない。そこで、パブリックアートと公共トイレが一緒になったらいいのではないか、と考えました。

では、「アートとトイレのハイブリッド」を目指すには、どんな表現がいいか。生活の中にありながら、人びとに常に何か問いかけてくる存在のあり方をさまざまに考え、最終的に、「79億通りのライティングパターン」に辿り着きました。79億というのは、2021年時点の世界人口と同じ数です。パターンをその数にしたのは、「THE TOKYO TOILET」のプロジェクト全体の根底に、「人は、みんな違うという意味で、同じである。」という思想があるからです。

ひとつのパターンが投影されるのはたった10秒。その後、15秒で次のパターンへゆっくりと変わります。最新の技術を使ってはいますが、テーマは普遍的で、受け取られ方の幅も広いもの。日常的に並木道を通る方々にとっては、毎日目にするその「変化」が、トイレとのゆるやかなコミュニケーションになってくれれば、と願っています。

The site is on a tree-lined street leading to Hiroo Garden Hills, with the University of the Sacred Heart building in the background. In designing the Hiroo East Park Public Toilet, we have always felt a responsibility for making it a part of the daily lives of its users, including the residents who have lived there for a long time.

First, we thought about how to make toilets more likable and cherished. Rather than creating toilets that people like simply because of their cleanliness, we wondered if we could create a system that would encourage people to like toilets more

南面。背後のライティングパターンを主役とし、建物そのものはできるだけシンプル、かつシャープな印象を目指した。屋根は鉄板。

Perspective view of the south elevation. The lighting patterns in the rear of the building are the project's main feature, and the building itself is designed as simple and sharp as possible. The roof is made of steel roof panels.

平面図。衛生設備機器の配置は左右対称で、2室のユニバーサル・トイレのうち、左にオストメイト配慮器具、右にベビーシートを備えた。

Floor plan. The facility has two universal toilet rooms with symmetrical sanitary equipment layouts. The room on the left is equipped with ostomate-friendly equipment, while the room on the right is fitted with a baby changing station.

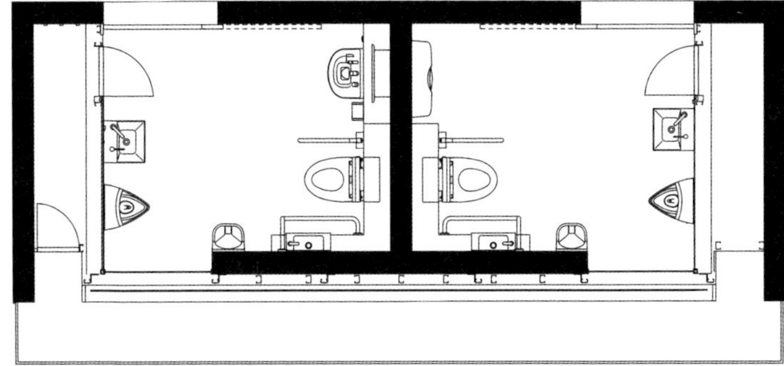

enthusiastically and create a new relationship with the city.

Public art is loved and cherished by everyone in most cases. While some people may spoil toilets, few people neglect public art. This is why we conceived a hybrid of public art and public toilet.

Then, what kind of expression would be best to achieve a "hybrid of art and toilet? After pondering various forms of existence in daily life that would always pose questions to people, we decided to make "7.9 billion different lighting patterns." 7.9 billion represents the world's population as of 2021. We decided to create the same number of patterns because the entire THE TOKYO TOILET project stands true to the idea that "people are all different, and therefore are all the same."

Each pattern appears for only 10 seconds. Then it slowly changes to the next pattern in 15 seconds. While it uses cutting-edge technology, the theme is universal, and people can perceive it in many different ways. For those passing by the tree-lined street daily, the changes observed every day will hopefully trigger gentle communication between people and the facility.

Drawings and Specifications ▶ P.264

後智仁 Tomohito Ushiro

アートディレクター／クリエイティブディレクター。1971年生まれ。博報堂を経て、2008年にWHITE DESIGNを設立。企業ブランディング、コミュニケーション設計、広告の企画・デザインを中心に、インテリアデザインやアートワークも行う。

Art director/creative director, born in 1971. After working at Hakuhodo, Tomohito Ushiro established WHITE DESIGN in 2008. His work centers on corporate branding, communication design, and advertising planning and design, along with interior design and artwork.

裏参道公衆トイレ

URASANDO PUBLIC TOILET

デザイン：マーク・ニューソン　渋谷区千駄ヶ谷4丁目28番1号

Design : Marc Newson　4-28-1 Sendagaya Shibuya-ku

2023年1月完成。鉄筋コンクリート造。建物は正方形で、建築面積は約45m²。湾曲した箕甲の納まりが美しい銅葺き屋根は、社寺建築を専門とする職人が手がけた。

Completed in January 2023. Reinforced concrete construction. The building is square in plan with a square footage area of approximately 45m². The copper roof, with its beautiful detailing of curved Minoko, was built by carpenters specializing in shrines and temples.

角はすべて曲面とし、内装は淡いブルーグリーンの単色でシームレスに仕上げられている。フラットな天井のダウンライトもマーク・ニューソンによるデザイン。

All corners are rounded, and the interior is seamlessly finished in a single hue of pale blue-green. Newson also designed downlighting in the flat ceiling.

コンクリート打ち放しのゲートも曲面仕上げ。高い施工技術が必要とされた。入口正面にユニバーサル、右に女性用、左に男性用トイレ。床にもピクトグラムを配した。

The concrete entryway also has rounded corners, requiring a cutting-edge construction technique. A universal toilet room is placed in front of the entrance, a women's toilet room on the right, and a men's toilet room on the left. Pictograms are placed on the floor as well as on the walls.

左：ユニバーサル・トイレ。内装の色調は施設全体で統一している。左中：男性用トイレの個室ドア付近。右中：光を柔らかく反射するブルーグリーンの床。右：外構照明は、屋根と外壁の間のくぼみに仕込まれている。

Far left: View of the universal toilet. The interior color tone is consistent throughout the whole facility. Center left: View near the door of the men's toilet cubicle. Center right: The blue-green floor softly reflects light. Far right: Exterior lighting is tucked into the recess between the roof and the exterior wall.

四方に傾斜する寄棟屋根のラインが目を引く外観夕景。敷地頭上には、首都高速道路4号新宿線が走る。石垣の立ち上がりは最も高いところで歩道から約1.5m。

The exterior evening view of the building, with eye-catching lines of the hipped roof sloping in four directions. The Metropolitan Expressway No. 4 Shinjuku line passes directly over the site. The stone wall rises roughly 1.5 m from the sidewalk at its highest point.

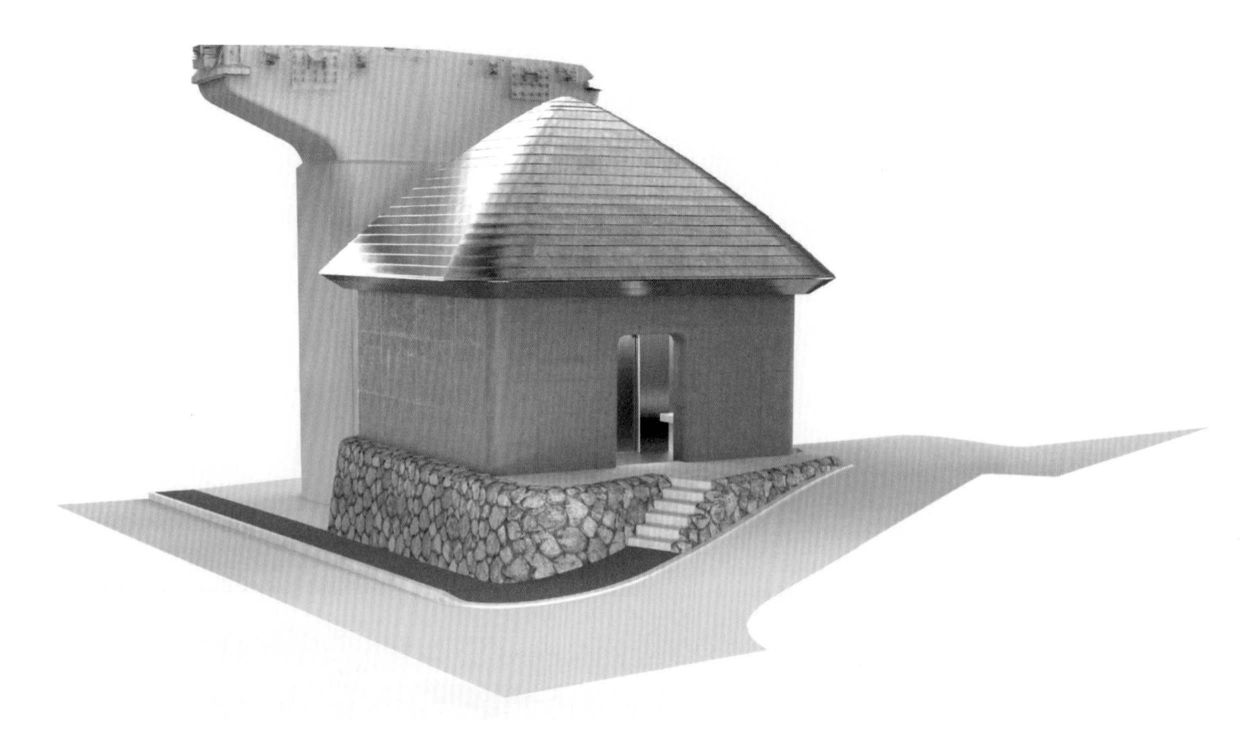

施設南東の完成イメージパース。「機能性、シンプルさ、そして心地よく永続的な空間であること」に重点を置いてデザインされた。

Perspective view of the southeast portion of the facility. The design focused on functionality, simplicity, and creating a comfortable and enduring space.

伝統的な日本建築の引用と、インダストリアルデザインの融合

Merging Traditional Japanese Architectural References with Industrial Design

「裏参道公衆トイレ」の立地条件は複雑でした。地形に高低差があり、上空は首都高速道路の高架に覆われています。また、公共の駐輪場も隣接しているなど、物理的に限定される条件を整理し、克服することが、デザインにおける重要なプロセスでした。

内部空間は、各要素が明確かつ具体的な機能をもっていることから、ひとつのプロダクトとして考えました。比較的狭いスペースの中に、満たすべき要件が与えられ、それらすべてを考慮する必要があるという点は、実際にプロダクトのデザインとよく似ています。

外部のデザインは、銅板の箕甲(みのこ)屋根をはじめとする日本の伝統的な建築の引用が中心となっています。賑やかで超近代的な場所にあっても、神社仏閣や茶室、農村部などによく見られるこの屋根の形が、潜在的に心地よさや安らぎを感じさせるものにしたいと思いました。源となったのは、私の長年の日本との親密な関係と、日本の職人技から多大なインスピレーションを受けてきたことにあります。屋根も石垣も、古くなるほど味が出て、長い年月を経たあかつきには、東京を織りなす風景の一部となることでしょう。

外部の素材では、コンクリートも私にとって日本らしい素材のひとつです。現代の日本では、コンクリートが非常に洗練された方法で使用されています。街なかのあらゆる建造物で、素晴らしいクオリティで使われている。日本以外の国と比較すると、そのクオリティの高さはとても特殊で、理想的でもあります。コンクリートはこの場所を永続的な空間にするための主構造であり、「裏参道公衆トイレ」が周囲に溶け込み、現代日本の風景の一部になるためにも適した素材だと思います。

「THE TOKYO TOILET」のようなプロジェクトは、現時点では、日本の東京でしか起こり得ないものでしょう。土地の権利者や自治体、民間団体や企業が協力し、チームをつくり、公共トイレの改善を長期的に主導するというのは、極めて珍しく、今後は他の都市の参考になると思います。利用者の皆さんには、ぜひこの空間に「歓迎されている」と感じてもらいたいです。

The location of the "Urasando Public Toilet" was complex. There is a considerable difference in elevation and the Metropolitan Expressway passes directly over the site. Sorting out and solving physical limitations of the site, including the adjoining public bicycle parking lot, was an essential part of the design process.

Newson's approach to the interior space was like designing

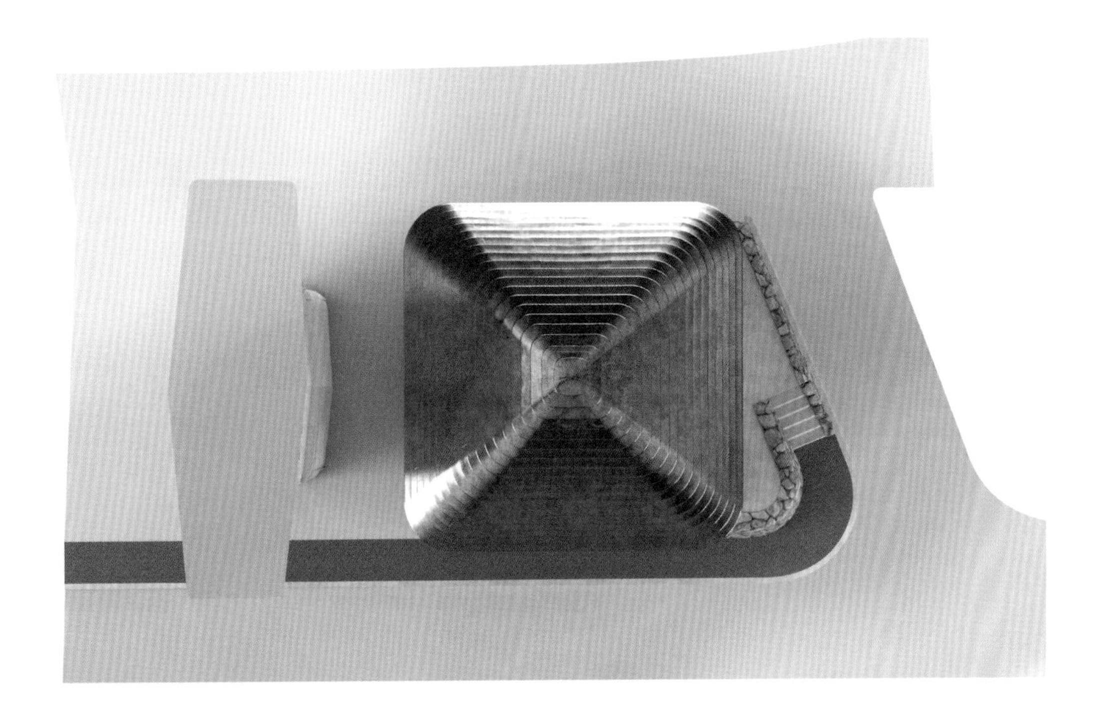

銅板の方形屋根の鳥瞰イメージ。日本の伝統的な建築を引用しつつ、全体に丸みをもたせ、できるだけ段差のない平坦な仕上げとした。

Bird's-eye view of the square copper roof. The roof, referencing traditional Japanese architecture, is rounded throughout and finished as flat and even as possible with no steps.

マーク・ニューソン Marc Newson

インダストリアルデザイナー。1963年生まれ。日用品からジェット機まで、ジャンルを超えた幅広い分野で活躍。クライアントにはAppleやエルメス、ナイキなどがあり、作品は世界各国の主要なデザインミュージアムにコレクションされている。ロンドン在住。

Industrial designer, born in 1963. His works span a wide range of genres from everyday items to jet planes. Clients include Apple, Hermes, and Nike, and his works are in collections at major design museums around the world. Based in London.

a product, given that each element needs to fulfill a clear and specific function in a relatively small space and serve a limited and definite purpose. In addition, public toilet spaces, like industrial design, consolidate various technological elements. Newson finds certain similarities between the interior design of public toilets and that of aircraft from his extensive experience in aircraft interior design.

The exterior design primarily draws from traditional Japanese architectural references, including the Minoko (copper sheet) roof. Newson wanted this roof shape, commonly found in shrines, temples, tea houses, and rural areas, to convey a sense of comfort and serenity even in a busy, ultra-modern place. The idea stems from Newson's longstanding close relationship with Japan and the great inspiration he has gained from Japanese craftsmanship. Roofs and stone walls become more attractive as they age and will eventually become a part of the landscape that weaves the fabric of Tokyo.

Concerning exterior materials, concrete is one of the materials he finds uniquely Japanese. It is used in highly sophisticated ways in modern Japan, in all kinds of buildings throughout the cities, with excellent quality. This high quality is very unique and even ideal compared to other countries outside of Japan. Concrete is the primary structure that makes this place a permanent space and a suitable material to blend the "Urasando Public Toilet" into its surroundings and to make it a part of the contemporary Japanese landscape.

A project like "THE TOKYO TOILET" could only happen in Tokyo, Japan, today. It is highly unusual to see a team of landowners, local government, private organizations, and businesses working together to take the lead in improving public toilets on a long-term basis. This project will serve as an excellent model for other cities to follow. Newson hopes that all users will feel welcome in this space.

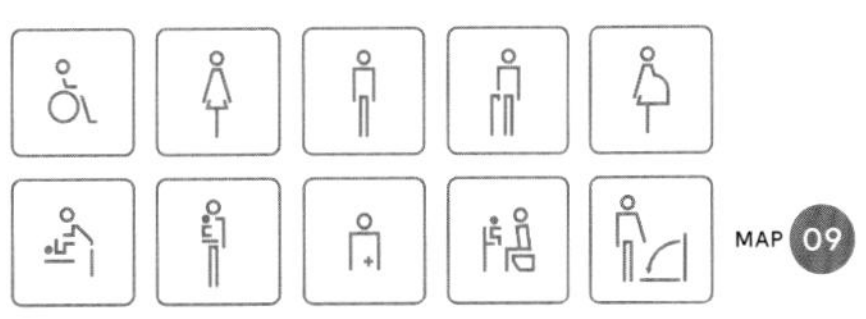

MAP 09

Drawings and Specifications ▶ P.266

2023年2月完成。鉄筋コンクリート造。施設中央のオープンスペースは幅6m。写真左手と正面にユニバーサル・トイレ、右手に男性用小便器スペースがある。

Completed in February 2023. Reinforced concrete construction. The open space in the center of the facility is 6 meters wide. The photo shows universal toilets on the left and center, and a men's urinal space on the right.

幡ヶ谷公衆トイレ

HATAGAYA PUBLIC TOILET

デザイン：マイルス・ペニントン／東京大学DLXデザインラボ

渋谷区幡ヶ谷3丁目37番8号

Design : Miles Pennington / UTokyo DLX Design Lab　3-37-8 Hatagaya Shibuya-ku

上：オープンスペースは展示などにも活用できるよう白壁を大きくとった。下左：人通りの多い中野通りと水道道路の交差点に面して建つ。下中：敷地南、水道道路側の側面。下右：敷地前面は歩道として整備した。

Top: The open space has large white walls suitable for exhibitions and other uses. Bottom left: The building stands at the busy Nakano Dori and Suido Doro intersection. Bottom center: View of the south side of the property, along "Suido Doro" (Aqueduct Road). Bottom right: The front of the site was developed as a walkway.

三角形の長辺一部を入口としたユニバーサル・トイレ内。天井高は最大で約4.7m。

Interior view of the universal toilet with an entrance on one of the long sides of the triangular building. The maximum ceiling height is about 4.7m.

男性用の小便器スペース内、手洗い場付近からエントランス方向を見る。
View toward the entrance from the hand wash basin in the men's urinal space.

左：天井の17灯のスポットライトがオープンスペースを照らす。右：敷地内から交差点を見る。床に埋め込んだ昇降式のポールを引き上げ、長さ1.2mの木柱をはめ込むとベンチに。何十通りもの組み合わせが可能。

Left: Seventeen spotlights on the ceiling illuminate the open space. Right: View of the intersection from the site. One can pull up elevating poles embedded in the floor and insert 1.2m-long wooden bars to create benches. They can be configured in multiple ways.

中野通り

街に対して開き、地域コミュニティの中心となることをイメージした鳥瞰パース。
Bird's eye view perspective of the building open to the city, envisioned as the center of the local community.

地域コミュニティの中心となる、第2の空間があるトイレ
Public Toilet with a Second Space as the Center of the Local Community

私たち「東京大学DLXデザインラボ」は、デザインとエンジニアリングの融合を目指し、国内外、さまざまな専門家とのコラボレーションで、革新的な製品やサービスのプロトタイプを提案しています。

「幡ヶ谷公衆トイレ」でも、まずは私をはじめ、デザイン先導イノベーションを専門とするチームがリサーチ活動を展開。多様な国籍、年齢、性別の方々とのワークショップを実施してアイデアを積み上げ、そのアイデアを建築専門のチームがカタチにしていくというコラボレーションでプロジェクトを進めました。地域住民の方々との対話を重ねたことも、「幡ヶ谷公衆トイレ」のデザインプロセスの特徴だと思います。

今回の提案の最も大きなポイントは、公共トイレに別の機能をもつ「第2の空間」を組み合わせたことです。その第2の空間は、間口の広いオープンスペースとし、年齢や性別にかかわらず、すべての人たちがさまざまな用途に活用できるよう、正方形のプレーンな空間としました。展示スペースや情報センターとしてなど、「地域コミュニティの中心」として役立てられることを期待しています。

オープンスペースの利便性を考え、昇降式のポールを使ってカタチを変えられるオリジナルのベンチも用意しました。ポールは全部で31個、床に埋め込まれています。休憩や待合い用のベンチとしてはもちろん、用途に合わせ、フォーメーションをその都度変えるなどして活用していただけるとうれしいです。

ユニバーサルと男女共用トイレ、そしてそれらの快適な利用を促すための男性用小便器のスペースは、敷地に残った3つの三角形の空間に分散して配置しました。広さは限られていますが、天井高と屋根の傾斜によって、面積以上の広がりを感じてもらえると思います。

We at the "UTokyo DLX Design Lab" strive to integrate design and engineering and propose prototypes of innovative products and services in collaboration with diverse experts from Japan and abroad.

In the Hatagaya Public Toilet project, a design-led innovation specialist team, including Miles Pennington, first conducted research activities. The project proceeded in collaboration, involving workshops with people of diverse nationalities, ages, and genders to develop ideas, which eventually took shape with the help of a team of architectural specialists. Extensive discussions with the community residents were one of the distinctive aspects of the design process of the Hatagaya

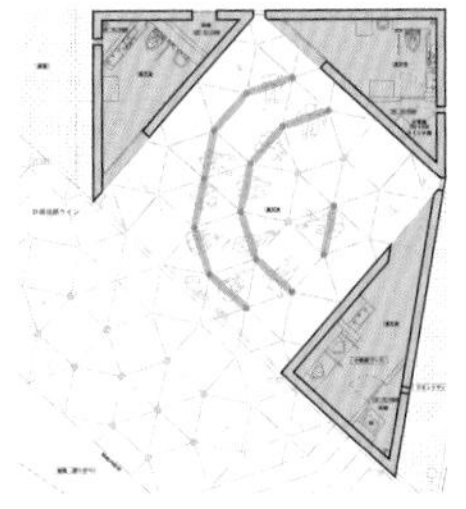

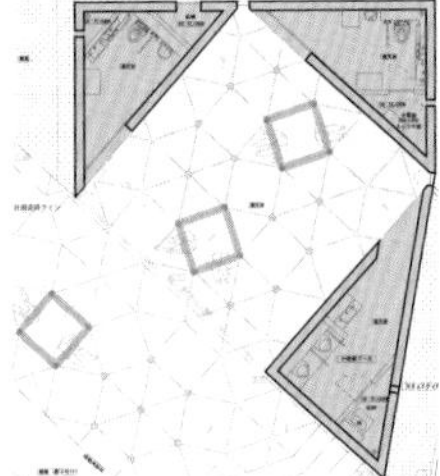

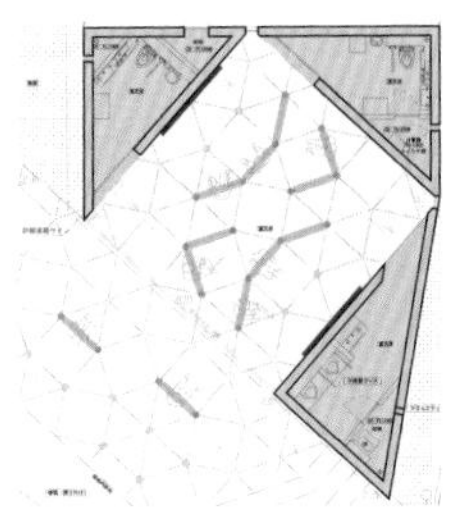

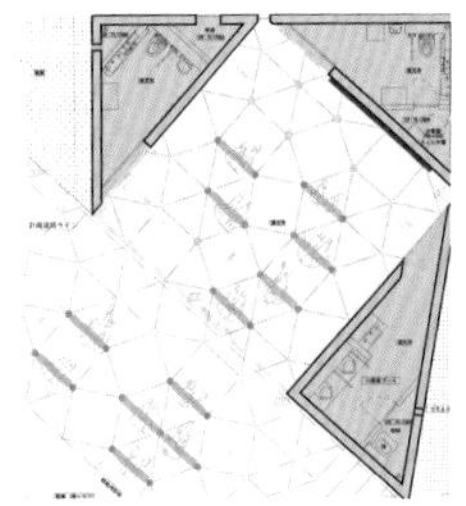

ベンチレイアウトのバリエーション例と、講演会や上映会などの利用イメージ。可変式のベンチシステムは、「幡ヶ谷公衆トイレ」の設備としてオリジナルで開発、デザインされた。

Examples of bench layout variations and images showing possible uses such as lectures and screenings. The variable bench system was developed and designed exclusively for the "Hatagaya Public Toilet.

Public Toilet.

Combining a "second space" with a separate function in a public toilet was one of the most significant aspects of this proposal. The second space is an open space with a wide frontage, plain and square, to allow all people, regardless of age or gender, to use it for various purposes. It is designed to serve as a center of the local community, for example, as an exhibition space or information center.

For the convenience of the open space, we also designed original benches with elevating poles allowing them to change shape. There are a total of 31 poles embedded in the floor. These benches may be used for resting and waiting and various purposes by changing the form as needed. The universal and unisex toilets, and the men's urinal space to facilitate more comfortable use of these facilities, are dispersed across the triangular corners remaining on the site. While the space is limited, the high ceilings and the sloping roof give a sense of spaciousness beyond the actual area.

Drawings and Specifications ▶P.268

マイルス・ペニントン Miles Pennington

研究者。英国ロイヤル・カレッジ・オブ・アートのイノベーション・デザイン・エンジニアリング、プログラム長などを経て、東京大学生産技術研究所デザイン先導イノベーション研究室、DLXデザインラボ教授。専門はデザイン先導イノベーション。

Professor at DLX Design Lab from IIS, The University of Tokyo, after serving as Director of Innovation Design Engineering at the Royal College of Art in the UK. He specializes in design-led innovation.

建築デザイン:
東京大学生産技術研究所
今井公太郎研究室
本間健太郎研究室

Architectural design: Kotaro Imai laboratory and Kentaro Honma laboratory from IIS, The University of Tokyo

笹塚緑道公衆トイレ

SASAZUKA GREENWAY PUBLIC TOILET

デザイン：小林純子　渋谷区笹塚1丁目29番

Design : Junko Kobayashi　1-29 Sasazuka　Shibuya-ku

2023年3月完成。耐候性鋼板のサンドイッチパネル造。中央にユニバーサル、左右に男性用と女性用トイレを振り分け、歩道側に子ども用トイレを設けた。敷地上は鉄道の高架。

Completed in March 2023. Weather-resistant steel sandwich panel construction. The universal toilet room is in the center, with men's and women's toilet rooms on either side and a children's toilet room facing the sidewalk. An elevated railroad runs over the site.

左上：ウサギのイラストはグラフィックデザイナーの太田徹也が手掛けた。左中、左下：男性用トイレ入口付近と手洗い場。右：女性用トイレ入口付近。壁は造船技術をもつ会社がパーツごとに製作し、現場で溶接した。

Top left: The rabbit illustration by graphic designer Tetsuya Ota. Middle left and bottom left: View of the men's toilet room entrance and the hand wash basin. Right: View around the entrance to the women's toilet room. A company specializing in shipbuilding technology fabricated the walls in parts and welded them together on-site.

左上：男性用トイレ内。左下：白が基調のユニバーサル・トイレ。右：手洗い場のカウンターをパウダー用に広くとった女性用トイレ内。乳白のガラス窓から自然光も入る。壁には断熱材を入れ、夏季の高熱と冬季の寒冷に備える。

Top left: Interior view of the men's toilet room. Bottom left: The universal toilet room featuring white as the primary color. Right: Interior view of the women's toilet room with a wide hand-washing counter for use as a powder room. Natural light enters through a milky-white glass window. The walls are insulated to withstand intense heat in the summer and cold in the winter.

左：子ども用トイレは、扉をあえて低くして、親が見守れるようになっている。右上：夜は黄色い大庇に光が反射し、高架下が明るく華やぐ。右下：背面歩道側は乳白のガラス窓を多用。中にいる人の気配が伝わる。

Left: The children's toilet room has a low door, allowing parents to watch over their children. Upper right: At night, the large yellow canopy reflects the light and brightens the area under the elevated railroad tracks. Lower right: The windows along the rear sidewalk are mostly milky-white glass. People can feel the presence of people in the building.

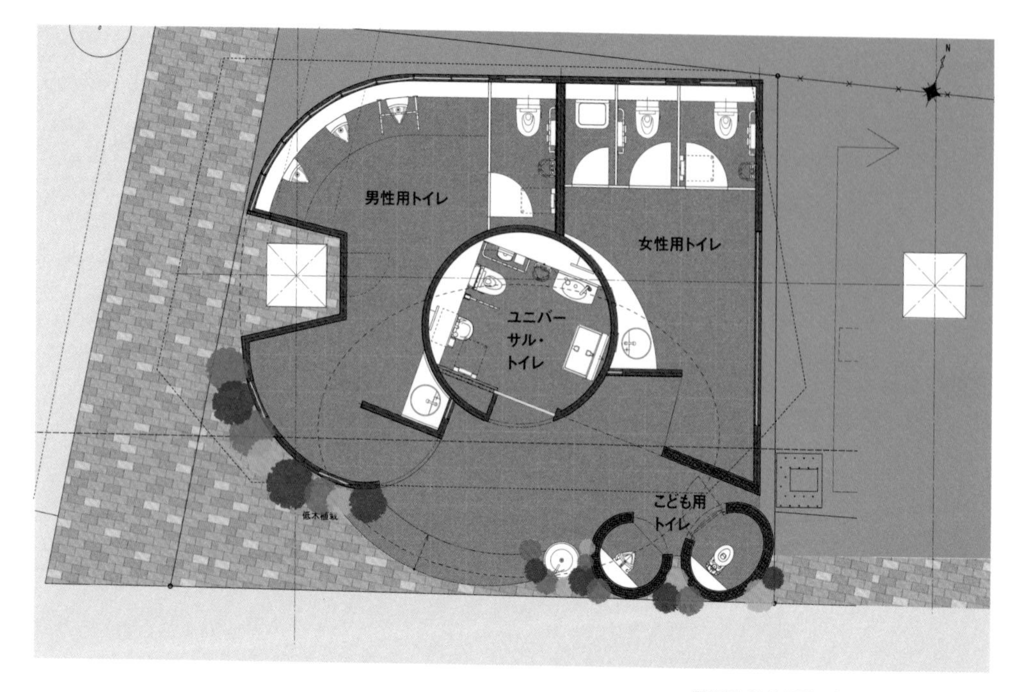

平面図。建築面積は約98m²。歩道に面した円形の子ども用トイレは、右が大便器、左が小便器。子ども用の手洗い場は個室外にある。

Floor plan. The building area is about 98 m². The children's toilet room is circular, facing the sidewalk, with a toilet on the right and a urinal on the left. The hand wash basin for children is outside the toilet room.

高架下に耐候性鋼板で立ち上げた、月とウサギのトイレ

Moon and Rabbit Toilet Facility Made of Weather-Resistant Steel Sheets under the Elevated Railroads

京王線の笹塚駅にほど近いこの場所は、鉄道の高架を支える橋脚とその基盤部分が地中にあり、また敷地南側には玉川上水の水路があることから、施工方法や荷重に制限がありました。建物の素材に耐候性鋼板パネルを選んだのは、大型のクレーンや重機を極力使わずに組み上げられ、重さを抑えられるからです。

鋼板には、自由な曲線が作れること、外壁はメンテナンスフリーであることなど利点もありました。建物の形そのものは、高架下という条件を撥ね除けつつ、市民に愛される形を目指しました。中央にユニバーサル、左右に男性用、女性用、そして歩道からすぐ入れる場所に子ども用トイレを配し、ゾーン毎に円筒の高さを変えることで、さまざまな利用者に対応するトイレが用意されていることを示せれば、と考えました。

トイレですから、使いやすさと安全性が第一です。そのうえで今回は、この場所ならではの工夫として、高架下の圧迫感を和らげるため、建物の上に「第2の空」として丸い大庇をかけました。黄色くしたことで、「お月様みたい」と言われることもあり、丸窓にウサギもいて、愛らしい月とウサギのトイレになったと思います。公共トイレは、利用者の身近な存在になってもらえることが大切だと考えています。施設に面した歩道は、近隣の保育園の散歩コースでもあります。子どもたちにもぜひ、よろこんで使ってもらいたいです。

私はこれまで、日本全国、250カ所以上の公共トイレの設計に携わってきましたが、公共トイレの実力は、30年、40年経って出るものだという実感があります。耐候性鋼板はわざと錆びさせた鋼板で、長い年月に耐えうる強度があると同時に、経年変化で黒味を帯び、より深みのある色になっていくのが魅力です。今はまだ溶接の跡などが見える「笹塚緑道公衆トイレ」も、時間と共に味わいが増し、この場所に馴染んでいくことでしょう。

「THE TOKYO TOILET」が、これまで4K（臭い、暗い、怖い、汚い）の代名詞となってしまっていた公共トイレの現状を解決しようとしていることは、社会的にも大きな意味をもつと感じています。このプロジェクトが、「公共トイレは、市民みんなの財産であること」をもう一度考え直すきっかけとなり、また、これを機にさまざまなトイレへの議論が展開されることを期待しています。

The site is located near Sasazuka Station on the Keio Line. It had restrictions on construction methods and loads because the piers and foundations of the elevated railroads are underground and the Tamagawa-josui water channel is located to the south of the site. Weather-resistant steel plate panels were chosen as the building material because they can be assembled with minimal use of large cranes or heavy machinery, minimizing the load.

Steel sheets have advantages such as the flexibility to create any curve and the ability to allow maintenance-free exterior walls. As for the shape of the structure itself, we sought a form

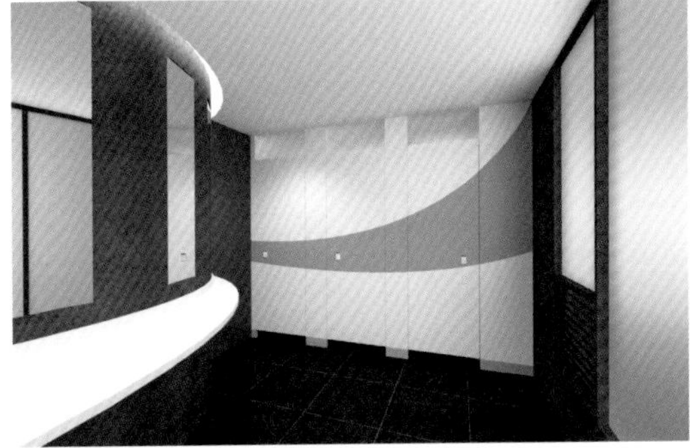

昼と夜の歩道からの見え方などを検討したイメージパース。建物の構造設計は、日本を代表する構造家の梅沢良三。右上のパースは女性用トイレで、奥まで入らなくても個室の空き状態がわかる見通しの良さを重視した。

Perspective drawings to study the building's visibility from the sidewalk during the day and at night. Ryozo Umezawa, one of Japan's leading structural engineers, undertook the structural design. The upper right perspective shows the women's toilet room, focusing on the visibility that allows users to check the occupancy of cubicles without going further into the space.

that citizens would cherish yet that would also overcome the condition imposed by its location under the elevated railroads. We wanted to show that there were toilet rooms for various users by placing a universal toilet room in the center, men's and women's toilet rooms on either side, and a children's toilet room just off the sidewalk, and by changing the height of the cylinders for each zone.

Ease of use and safety are top priorities for toilet facilities. In addition, as a unique feature for this location, a large circular canopy was placed over the facility as a "second sky" to alleviate the oppressive feeling under the elevated structure. The yellow color gives it a moon-like appearance, and with the round windows decorated with rabbits, it has become an adorable moon-and-rabbit toilet. We believe it is important for public toilets to be familiar to users. The sidewalk facing the facility is a walking course for nearby nursery schools, and we hope that the children will enjoy using the facility.

I have designed more than 250 public toilet facilities throughout Japan, and I feel that public toilets will reach their full potential after thirty or forty years. Weather-resistant steel sheets are steel plates rusted on purpose, strong enough to withstand long periods of time, becoming darker and more profound as they age. The Sasazuka Greenway Public Toilet still shows welding marks, but it will gradually harmonize with the place over time.

THE TOKYO TOILET is a socially significant project that aims to solve the current situation of public toilets, which until now have been synonymous with 4K, namely "kusai, kurai, kowai, kitanai" (smelly, dark, scary, and dirty). We hope the project will inspire people to reconsider the idea of public toilets as an asset for all citizens and stimulate further discussions on various issues surrounding toilets.

Drawings and Specifications ▶ P.270

小林純子 Junko Kobayashi

建築家。ゴンドラ代表。一般社団法人日本トイレ協会会長。1988年に5億円をかけた香川県の公共トイレ「チャームステーション」に携わって以来「笑顔の生まれる快適なトイレ」の設計をライフワークに徹底した利用者目線でのトイレづくりを続ける。

Architect. CEO of Gondola. Chairperson of the Japan Toilet Association. Junko Kobayashi has been involved in designing "comfortable toilets that bring smiles to people's faces" as her life's work since 1988, when she was involved in the 500-million-yen "Charm Station" public toilet project in Kagawa Prefecture and continues to design toilet facilities from the users' perspective.

2023年3月完成。鉄筋コンクリート造。歩道から直接入れる敷地北側にユニバーサル・トイレ。男性用、女性用トイレはアーチ状のゲートから。通路は見通しがよく通り抜けできる。

Completed in March 2023. Reinforced concrete construction. The universal toilet room is on the north side of the site, directly accessible from the sidewalk. Men's toilet room and women's toilet rooms are accessible through the arched gate. The passageway is clearly visible and allows passage through.

西参道公衆トイレ

NISHISANDO PUBLIC TOILET

デザイン：藤本壮介　渋谷区代々木3丁目27番1号

Design : Sou Fujimoto　3-27-1 Yoyogi Shibuya-ku

左：歩道側のスロープに設置された蛇口。グースネックで向きは自由に変えられる。水はスロープを伝い、中央の排水溝へ。右：まばゆいばかりの白さと有機的な曲線に包まれる通路。

Left: View of faucets installed on the slope on the sidewalk side. The direction of each faucet can be changed freely by turning the gooseneck. Water runs down the slope to a draining ditch in the center. Right: The passageway is enveloped in dazzling white and organic curves.

左：女性用トイレのドア付近から街を見る。外壁には防汚性に優れた光触媒塗装が施されている。右：スロープの一角に窪みを設けシンボルツリーを植えた。

Left: View of the city from near the door of the women's toilet room. The exterior walls are coated with a stain-resistant photocatalytic coating. Right: Part of the slope is hollowed out to plant a symbolic tree.

左上：ユニバーサル・トイレ内。左下：敷地対向の歩道からの夜の外観。右：白壁の効果で各トイレの個室入口全体が明るく、夜は光のなかに入っていくような安心感がある。

Top left: Interior view of the universal toilet room. Bottom left: Night view from the sidewalk across the street from the site. Right: The white walls brighten up the entrance to each toilet room, giving a sense of security at night, as if entering into the light.

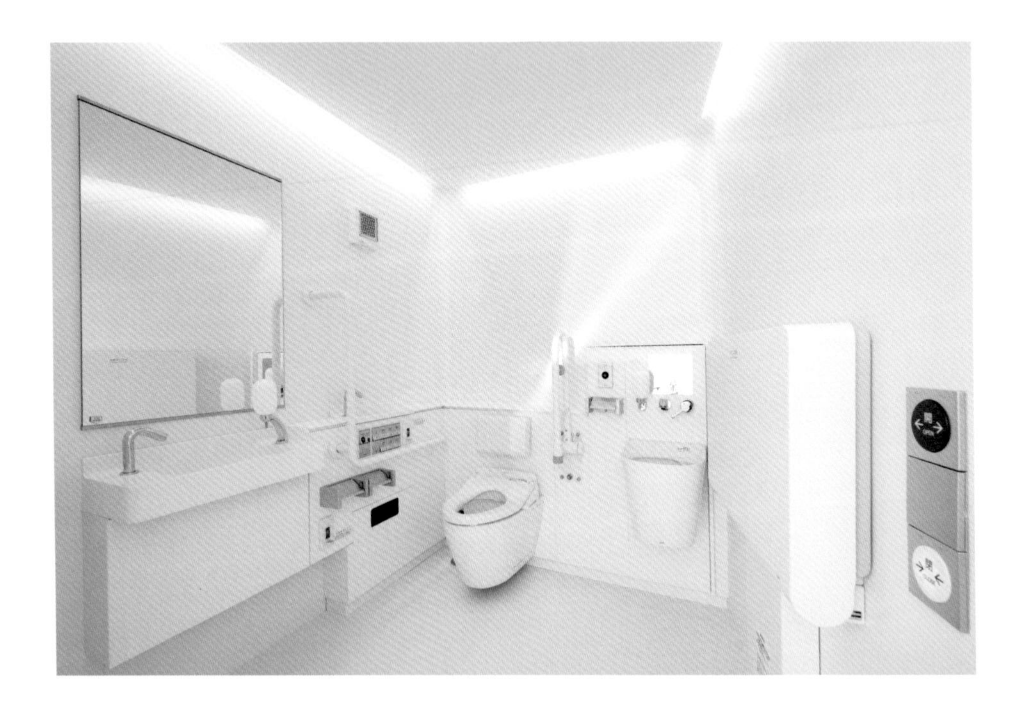

清らかな水の流れを思わせる曲線と、水場に集まる人びとの姿が描かれた、藤本壮介の「器・泉」のスケッチ。

Sou Fujimoto's sketch of "vessel/fountain" depicts a curved line evoking the clear flow of water and people gathering at a place with water.

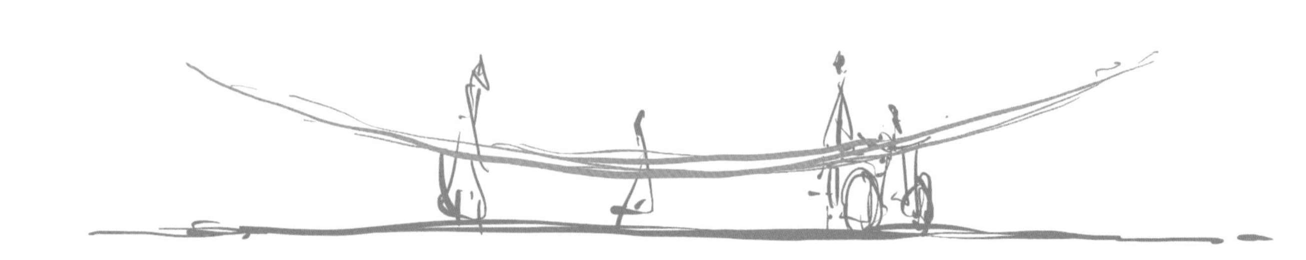

都市のなかに泉をもたらす、みんなのための、ひとつの器

A Vessel for All, a Fountain in the City

公共トイレは建物の規模が小さく、機能も限られています。そこに新たな可能性を見出していくプロセスはとても面白く、また難しい道のりでもありました。トイレであると同時に「街の価値」となるような、新しい公共空間をつくりたい——。その思いで、街とトイレとの関係を幾通りも考え、数年かけて辿り着いたのが「器・泉」というコンセプトです。

トイレは建築用語でいうところの「水まわり」です。そして水は、自然環境や循環の象徴でもあります。その「水のある場所」がトイレであると捉え直すと、今までの公共トイレとはまた違う役割や魅力を生み出せるのではないか、と思い至りました。

ヨーロッパの街なかでは、多様な人びとが集い、市民に親しまれている「泉のある広場」をよく見かけます。水場があると、人はそこに意識を向けるし、井戸端会議のように話も弾む。コロナ禍で手洗いの習慣が私たちにより根付いたことも、水場という存在の大切さに改めて気付いた要因かもしれません。

「西参道公衆トイレ」では、施設内からも歩道側からも使える5つの蛇口を設け、子どもから背の高い大人まで、それぞれが好きな場所を選んで、清潔な水にアクセスできるようにしました。中央が大きく凹んだ建物は、その高さの異なる手洗い場をひとつの形に内包したもので、みんなのための、ひとつの「器」です。

トイレとしての機能は、建替え前の施設同様、女性用の個室2室、男性用の個室1室と小便器3つを備えつつ、新たにユニバーサル・トイレを1室設けました。機能は増えましたが、以前より広く感じるのは、施設全体、つまり「器」の、歩道に対する開放感からだと思います。アプローチから人の往来が見え、通り抜けもでき、守られているという安心感もある。そのアプローチの塩梅にも気を配って設計しました。

これまでの公共トイレは「隠されている」ことが多かったように思います。存在が目立たないように隠され、使うとき以外は避けられていた。でも、みんなに必要なものなのだから、姿をもっと現していいいし、場所としてもっと開かれていい。街にあるものは、どんなに小さなものでも、それが良くデザインされていると場所も人も楽しく、豊かになります。街に対して開き、「器・泉」としてデザインした「西参道公衆トイレ」が、この場所をより良く、豊かにしてくれることを願っています。

Public toilets are small-scale buildings with limited functions. The process of finding new possibilities for this type of facility was very interesting yet challenging at the same time. We wanted to create a public space that would become a "value to the city" while serving as a toilet facility. With this in mind, we contemplated about the relationship between the city and the public toilet in many different ways and over the course of several years arrived at the concept of the "vessel/fountain."

Toilets are what we call "plumbing" in architectural terms. Water is also a symbol of the natural environment and circulation. By redefining the toilets as “places with water,” we concluded that toilets could play a different role and generate

さまざまな角度から「器」の見え方を検討したイメージパース。建物の高さはあえて抑え、3mほどとした。全体の面積は約40m²。

A perspective rendering to study the view of the "vessel" from various angles. The height of the building was deliberately kept low to about 3 m. The total area is roughly 40m².

attractive qualities from conventional public toilets.

In European cities, we often see fountain squares where a diverse range of people gather and are a familiar sight for the local citizens. A place with water attracts our attention and spark lively conversation like street corner gossip.

The fact that the COVID pandemic has ingrained in us the habit of hand washing may be another factor in our renewed awareness of the importance of places with water.

The "Nishisando Public Toilet" has five faucets that can be used both from inside the facility and from the sidewalk, allowing everyone from small children to tall adults to access clean water at their preferred spots. The building, with its deep concaved center, encapsulates the hand wash basins of varying heights into a single form, a single "vessel" for everyone.

The facility features two women's toilet rooms, a men's toilet room, and three urinals, the same as before the reconstruction, and a new universal toilet room was added. Even with the added features, the facility feels more spacious than before, due to the openness of the entire facility, or "vessel," to the sidewalks. The approach allows the users to see people coming and going, let them pass through, and provide a sense of security and protection. Our design carefully sought to balance these aspects.

Public In the past, public toilets have often been hidden away from view to minimize their presence and were avoided except when they needed to be used. But they are necessary for everyone, so they should be more visible and open as a place. No matter how small something is, if they are well-designed, they can bring joy and enrich the place and people. We hope that the "Nishisando Public Toilet," which we designed as a "vessel and fountain" open to the city, will improve and enrich this place.

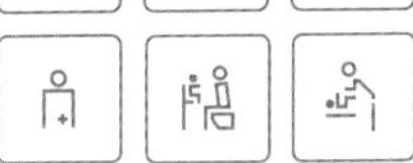

Drawings and Specifications ▶P.272

藤本壮介 Sou Fujimoto

建築家。1971年生まれ。2000年藤本壮介建築設計事務所設立。代表作に「武蔵野美術大学 美術館・図書館」など。パリ事務所を開設するなど海外でも活躍の幅を広げている。2025年日本国際博覧会(大阪・関西万博)の会場デザインプロデューサーに就任。

Architect, born in 1971, established Sousuke Fujimoto Architects in 2000. His representative works include the Musashino Art University Museum and Library. He is also expanding his activities overseas with the opening of his Paris office. He was appointed Venue Design Producer for the EXPO 2025 Osaka, Kansai, Japan.

PROJECT DESIGN

多様な利用者と建築に寄り添う、オリジナルのピクトグラム
Original Pictograms to Accompany Diverse Users and Buildings

佐藤可士和
Kashiwa Sato

「THE TOKYO TOILET」のピクトグラム開発にあたっては、日本の首都・東京に設置される新たな公共トイレのスタンダードとなるグラフィック・シンボルを目指しました。16組の多彩なジャンルのクリエイターが公共トイレをデザインするという企画意図から、建築デザインも多様なものになることが当初から予想されたため、その個性豊かな建物にも親和性の高いピクトグラムになるよう心がけました。

具体的にまず重視したのが、認識のしやすさです。国の産業規格であるJISを参考に、エレメントを整理しシンプルにすることで、視認性を高めています。また一つひとつのパーツの角を少しだけ丸くすることで、ユーザーに親しみやすい印象を与えられるよう意識しています。視認性と親しみやすさというこの2つのポイントを押さえ、プロジェクトを象徴するような存在感を併せ持つデザインとしています。

「THE TOKYO TOILET」を訪れる世界中の多様な文化の人びとに瞬時に認識されるよう、文字なしでも判断できることはデザインの重要なポイントだと考えていました。ただ、文字との併用の有無は、ルールとして定めているわけではありません。結果的に、17の建物すべてで、ピクトグラムのみの表記となり、色分けもされませんでしたが、それらはすべて、各クリエイターの判断にお任せをしています。素材は耐久性を重視し、ステンレスを基本とする案も検討しましたが、最終的には素材も、それぞれの建築デザインに合わせて自由に選択していただいています。

ベビーチェアや介助用ベッドなど、これまでの公共トイレではあまり見かけなかったであろうピクトグラムも数多く開発しました。プロジェクトの趣旨とバリエーション豊かな立地環境を鑑みると、これも当初から、多種多様なトイレの機能が盛り込まれることが予想されたため、必要と思われるピクトグラムはできる限り用意する意向をプロジェクト発案・資金提供者や日本財団に伝え、全12種類となりました。他に、掃除用具入れの場所やドアの開閉方法を示すサインも開発しています。

障がいのある方が使える設備
Accessible facility

女性
Women

男性
Men

高齢者優先設備
Priority facilities for elderly people

妊娠中の方優先設備
Priority for expecting mothers

ベビーシート
Baby changing station

乳幼児連れ優先設備
Priority for those with small children

子ども用トイレ
Children's toilet

オストメイト用設備
Facilities for ostomy

介助用ベッド
Care bed

ベビーチェア
Baby seat

着替え台
Changing board

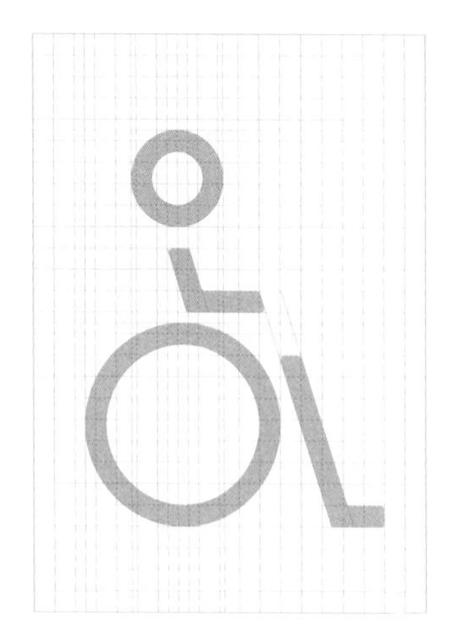

デザインをするなかで気づいたのは、既存の公共トイレのピクトグラムは、たとえば女性、男性などの識別のみを表すものと、障がいのある方が使える設備を示すものとでは、デザインエレメントが統一されていないケースが多かったことです。女性、男性だけのシンプルな識別がベースで、多様な利用者への配慮が見られない表記は好ましくありません。今回は、すべてを同一のルールで表現することを前提に、全体のデザインの方向性を定めていきました。ひとつのピクトグラムの中に複数の人物がいる場合、それぞれの関係性を表すことが難しくはありましたが、一方の人物を線のみで描き、もう一方の人物の頭部をベタ塗りにするという、ネガポジの表現で煩雑さを回避しています。

使用する際の規定ルールはひとつだけ。線幅はピクトグラムのサイズに比例し、プロポーションを変更しないこと。使用サイズも並べ方も、各クリエイターにお任せしています。立体化や発光化も容易にできるシンプルさ、縦に並んでも横に並んでも、小さいサイズで使用しても、ユーザーに認識されやすいアイコニックな印象。それが、このピクトグラムのデザインの強度です。それぞれのトイレのコンセプトに寄り添いながら、「THE TOKYO TOILET」としての統一感を生み出すことに寄与できたのではないかと考えています。

In developing the pictograms for THE TOKYO TOILET project, we aimed to create graphic symbols that would set a new standard for public toilets in Japan's capital, Tokyo. Given the project's intent to involve sixteen groups of creators from diverse genres in designing public toilets, we anticipated from the outset that the architectural design would also vary. Accordingly, we sought to create pictograms that would be highly compatible with the unique characteristics of these buildings.

The first thing we specifically focused on was ease of recognition. Using the Japan Industrial Standard (JIS) as a reference, we have organized and simplified the elements to enhance visibility. The corners of each part are slightly rounded to give a friendly impression to the users. The design is based on two key points, visibility and approachability, and exerts a strong presence as a symbol of the project.

Considering the diverse cultural backgrounds of people from all over the world who would be using THE TOKYO TOILET's facilities, we thought it was important for the design to be instantly recognizable without text. However, we did not stipulate using pictograms in conjunction with text. The decision

「THE TOKYO TOILET」オリジナルのピクトグラムのデザインプラン。左から、障がいのある方が使える設備、ベビーチェア、妊娠中の方優先設備。複数人いる場合でも同一のルールを用いて示すべき要素を単純化し、視認性を高めた。

Design plan for the original pictograms of THE TOKYO TOILET. From left to right: facilities for people with disabilities, baby seats, and facilities prioritizing pregnant women. The same rules were used to simplify the elements and increase visibility, even when multiple persons are in a single pictogram.

was up to each creator, and as a result, all seventeen buildings were marked only with pictograms without color coding. We also considered the idea of using stainless steel as the primary material, with an emphasis on durability. Ultimately, however, we let each creator choose a material that best suits their architectural design.

We have also developed several pictograms, such as baby seats and care beds for caregivers, many of which would not have been commonly seen in public toilets in the past. Considering the purpose of the project and the diverse locations, we anticipated from the outset that the facilities would incorporate many different types of toilet functions. We informed the project initiator/ sponsor and The Nippon Foundation that we would provide as many pictograms as we deemed necessary, resulting in twelve types of pictograms. In addition, we developed other signs, including the location of custodial closets and how to open and close the doors.

During the design process, we noticed that in many cases, the design elements of existing pictograms for public restrooms were inconsistent, for example, between those that only identify women and men, and those that indicate facilities for people with disabilities It is not desirable to have signage based on simple identification of only women and men, without consideration for diverse users. For this project, we set the overall design direction on the premise that the pictograms should express everything under the same rules. This direction, however, made it difficult to show the relationship between the multiple figures in a single pictogram. So, we avoided the complication by using negative-positive inversion: one figure was drawn with only lines while the head of the other figure was painted in solid color.

Only one rule governs the use of the pictograms. The line width must be proportional to the size of the pictogram and must not change proportions. The sizes and arrangements are left up to each creator. The strength of this pictogram design is its simplicity, which can easily be made three-dimensional or luminous, and its iconic impression, easily recognized by users, whether used vertically, horizontally, or in small sizes. We believe that our design contributed to giving a sense of unity to THE TOKYO TOILET project while staying close to the concept of each toilet facility.

清掃員は専属で、2人1組で各施設を回る。
ユニフォームのデザイン監修はNIGO®。

Dedicated cleaners, working in pairs, visit each facility. The uniform design was supervised by NIGO®.

THE TOKYO TOILETのメンテナンスについて

Maintenance of THE TOKYO TOILET

「THE TOKYO TOILET」は、プロジェクト発足当初から完成した施設を「きれいに使い続けること」を大切に考え、「メンテナンスの仕組みのデザイン」を模索してきました。

紺色のワークウェアにロゴを印刷したオリジナルのユニフォームは、デザイナーのNIGO®さんに監修をお願いしました。清掃員は、安全で清潔な環境づくりのプロフェッショナル。オリジナルのユニフォームを用意することで、日常のメンテナンスに注目が集まり、自分たちが使うトイレをきれいにしてくれる方々への感謝や敬意につながれば、との思いがありました。実際に着て作業をするスタッフからも好評で、「THE TOKYO TOILET」の清掃員であることが一目でわかるため、安心して声をかけてもらえることも増えました。

維持管理の運営は、日本財団と渋谷区、そして渋谷区観光協会の三者で取り組んできました。

まず、「THE TOKYO TOILET」の清掃には、3種類あります。必要に応じて1日3回行う通常清掃、1カ月に1回行う定期清掃、そして、年に1回の特別清掃です。清掃員は、公共施設の清掃業務を数多く手がけてきた「東京サニテイション」に依頼しています。

さらに、第三者機関である「トイレ診断士」による各トイレの現状調査と診断を実施。月に1回の定期診断と、年に1回の特別診断があります。「トイレ診断士」は、トイレのメンテナンス専門の関連企業で構成された「アメニティネットワーク」が社内検定を行い、認定している資格で、厚生労働省認定です。

この両団体・企業からの清掃報告、診断報告を元に、毎月、関係者による維持管理協議会を開催。トイレの利用状況を把握し、運用しながら維持管理業務の改善を図ってきました。このように、複数の組織・企業によって結成されたメンテナンスチームが公共トイレの維持管理にあたる仕組みそのものが新しく、この仕組みは2024年3月末に渋谷区に維持管理業務を移管するまで続ける予定です。

建物や設備を傷めない、丁寧な清掃活動

3種類の清掃は、それぞれ特長があります。1日1〜3回の通常清掃は原則「乾式」です。あまり水を使わない乾拭き中心の清掃方法で、手間も時間もかかりますが、菌の繁殖が抑えられるなどのメリットが多く、それぞれの場所の汚れや材質に適した専用の除菌・消臭剤とダスターなどを使い、手作業での拭き取り作業を行っています。鏡の水垢まで一つひとつ拭き上げるという、ホテルの客室さながらの丁寧な掃除は、施設を

清掃中
ご協力お願い
いたします。

「恵比寿東公園トイレ」にて。1日3回の場合、朝6時台から5〜6時間おきに清掃が入る。
At Ebisu Higashi Park Public Toilet, cleaning is performed three times a day, starting around 6:00 a.m. every five to six hours.

できるだけ傷めず、長く、きれいに使ってもらいたいから。月に1回の定期清掃は「湿式」で、専用の溶剤を使用し、通常清掃で落としきれず溜まった水垢、黒ずみ、全体の黄ばみなどを除去。年に1回の特別清掃は、外壁、照明設備、屋根の上の落ち葉除去など、見えないところまでくまなく清掃します。

清掃員からは、こんなメッセージも受け取っています。「THE TOKYO TOILETは建物ごとに素材や仕上げが異なるため、清掃方法の工夫も一様にはいきませんが、個性豊かな空間には、それぞれ愛着も生まれています。公共トイレは清掃を怠ると、どんどん汚れが蓄積してしまいますが、人は、美しいものは汚しづらいものです。毎日きれいな状態を維持することで、利用者に次の利用者のことを思って使っていただけることを信じています」

使い続けながらメンテナンスの向上を目指す

月に1回行われる「トイレ診断士」による診断項目は、汚れの指標となるATP（アデノシン三リン酸）拭き取り検査から換気能力の測定、衛生設備機器や非常用押しボタンなどの動作確認まで、多岐にわたります。専門家の立場から科学的な分析を行い、不良の有無を細かく調べ、快適度を数値化した「トイレ診断報告書」が、メンテナンス計画の道標。年に1回の特別診断では、アンモニア濃度測定や個室内の照度、温湿度の測定も行い、継続的な利用についてのさまざまなアドバイスも受けています。

診断報告書からわかっていることのひとつに、汚れを数値化したATP値が、供用開始当初より下がっているトイレもあるという、望ましい動向があります。ATP値の下降は、それぞれの施設に最適な清掃のアプローチを見つけ出し、日々、丁寧な掃除を行ってきた結果だと考えています。逆にATP値の上昇を確認した場合は、ダスターの種類を変え、拭き上げ方を工夫するなど、試行錯誤を繰り返しながら利用環境の改善を目指しています。

清潔で快適な公共トイレを維持するには、多くの人の力が必要です。施設を作るだけではなく、よりよく使う、みんなで守る。きれいな状態を維持することで、きれいに使われる、という好循環が生まれ、「公共トイレはきれいに使うのがあたりまえ」という心の醸成が図られていくこと。それが、「THE TOKYO TOILET」がこれからも目指していく、大きな目的のひとつです。

The Tokyo Toilet

通常清掃の様子。専用の除菌・消臭剤とダスターを使った「乾式清掃」で、便座の裏まで徹底して拭く。丁寧な日常清掃が建物や設備の劣化を防ぎ、利用者の快適と安全を支える。

Regular cleaning involved wiping down to the back of the toilet seat using "dry methods" with special sanitizers, deodorizers, and dusters. Careful daily cleaning prevents deterioration of the facility and equipment and provide comfort and safety for the users.

「笹塚緑道公衆トイレ」にて、「トイレ診断士」による定期診断の様子。設備の不具合や汚れ、換気量などを専門の機器を用いて調べ、報告書を作成。汚れの拭き取りや除菌なども行う。

Periodic assessment by toilet assessors at Sasazuka Greenway Public Toilet. They inspect equipment malfunctions, contamination, and ventilation levels using specialized equipment and prepare an assessment report. They also wipe away dirt and sanitize the facility.

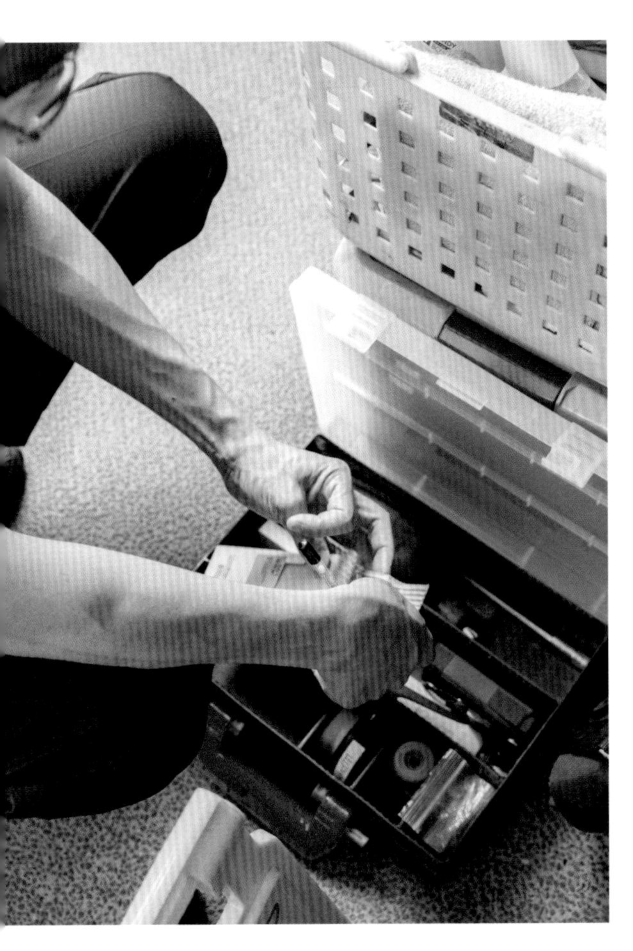

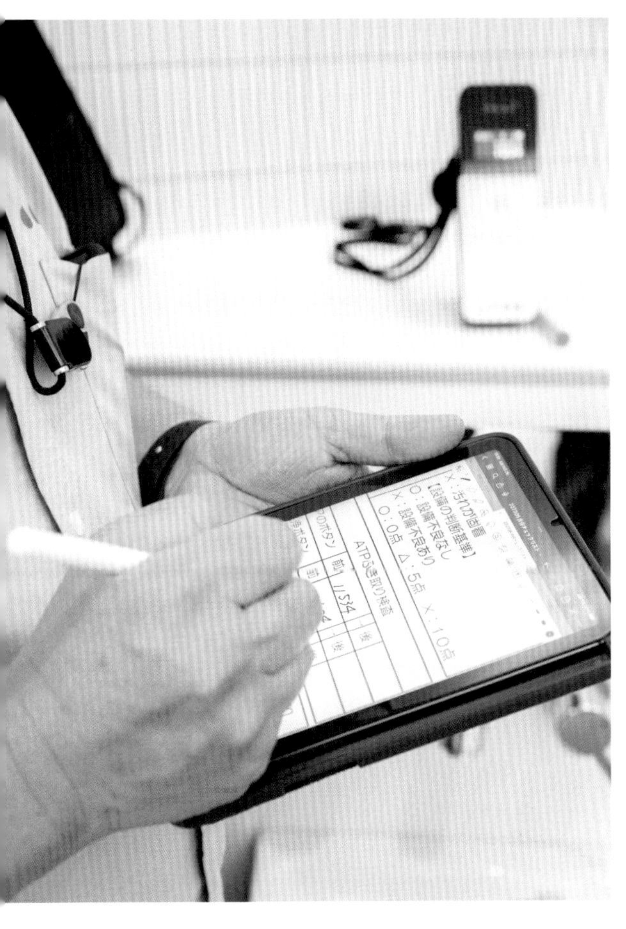

Since its inception, THE TOKYO TOILET has sought to design a maintenance framework to achieve continued clean use of the completed facilities.

The original uniforms, with logos printed on navy blue workwear, were supervised by designer NIGO®. The cleaners are professionals responsible for creating a safe and clean environment. We hoped that providing original uniforms would draw attention to their daily maintenance and lead to appreciation and respect for those cleaning the toilets they use. The uniforms have been well received by the staff who actually wear and work in them. Because users can tell at a glance that they are the cleaners for THE TOKYO TOILET, they feel more at ease and often interact with them.

Maintenance and management operations have been a three-way effort between The Nippon Foundation, Shibuya Ward, and the Shibuya City Tourism Association.

First, cleaners perform three types of cleaning at THE TOKYO TOILET facilities: regular cleaning which is carried out three times a day as needed; periodic cleaning which is carried out once a month; and special cleaning which is carried out once a year. The cleaners are hired by Tokyo Sanitation, a company with extensive experience in cleaning public facilities.

In addition, a third-party "Toilet Assessor" conducts surveys and assessments of the current condition of each facility, with monthly assessments and a special annual assessment. The Toilet Assessors are certified by the Amenity Network, an organization of companies specializing in toilet maintenance, which conducts in-house examinations and certification, and is recognized by the Ministry of Health, Labour and Welfare.

The concerned parties hold monthly maintenance and management meetings based on the cleaning and diagnostic reports from the organizations and companies. They work to improve maintenance and management operations while monitoring and operating toilet use. This system of maintenance teams formed by multiple organizations and companies to undertake maintenance of public toilets is new in itself, and will continue until the end of March 2024, when maintenance and management operations will be transferred to Shibuya Ward.

Careful cleaning operations that do not compromise buildings and equipment

Each of the three types of cleaning has specific characteristics. Regular cleaning which takes place one to three times daily generally uses "dry" methods. This cleaning method mainly involves dry wiping without water. While it is time-consuming and labor-intensive, it has many advantages, such as preventing the spread of bacteria. Careful cleaning, similar to cleaning a hotel room, includes wiping water stains off mirrors one by one and is intended to keep the facilities clean for as long as possible without compromising the facilities. Monthly cleaning uses "wet" methods, in which cleaners use special solvents to remove accumulated water stains, dark stains, and overall yellowish discoloration that cannot be removed by regular cleaning. Special cleaning performed once a year covers areas that generally receive little attention, such as exterior walls, lighting fixtures, and rooftops, and includes the removal of fallen leaves on the roof.

We have received a message from the cleaners as follows:

"Each of THE TOKYO TOILET facilities is built with different materials and finishes, so cleaning methods vary from facility to facility. Each facility has a unique space, and we have grown attached to them. If public toilets are not cleaned properly, dirt will accumulate. People do not like soiling beautiful things. We believe maintaining cleanliness every day will encourage users to use the toilets with the next user in mind."

Striving to improve maintenance through continued use

Monthly assessment items performed by toilet assessors range from ATP (adenosine triphosphate) tests indicating contamination to measuring ventilation capacity and checking the operation of sanitary equipment, emergency buzzers, and other equipment. The Toilet Assessment Report, a detailed inspection of defects and quantification of comfort level based on scientific analysis by experts, is the guidepost for the maintenance plan. The annual special assessment includes measurement of ammonia concentration, illuminance, temperature and humidity in the rooms, and various advice on continued use of the facilities.

One of the things we learned from the assessment report is that the ATP value, which quantifies contamination, has dropped in some facilities since they were first put into service, which is a desirable trend. We believe the decline in ATP values resulted from our efforts to determine the best cleaning approach for each facility and careful cleaning on a day-to-day basis. Conversely, if an increase in ATP value is observed, we aim to improve the toilet environment through trial and error, for example, by changing the type of duster and the wiping method.

Maintaining clean and comfortable public toilets requires the efforts of many people. Not only do we need to build facilities, but we need to use them better and maintain them together. One of the primary goals of THE TOKYO TOILET is to create a positive cycle in which maintaining cleanliness encourages clean use and instills a mindset that "public toilets should be always used cleanly."

CONCEPTION AND CONSTRUCTION

気づきを生み出し、心を動かす、THE TOKYO TOILET

柳井康治
THE TOKYO TOILET プロジェクト発案・資金提供者

誰もが快適に使える新しい公共トイレを東京につくりたい。その思いに至るまでには、さまざまな人との出会いや個人的な体験、そこからの気づきの蓄積がありました。

車いすテニス男子シングルスで史上初のグランドスラムを達成した、国枝慎吾さんとの出会いもそのひとつです。国枝さんを初めて知ったのは2008年のことですが、車いすテニスのカテゴリーを超え、かのロジャー・フェデラーからも日本のテニス界には国枝あり、と賞賛される強さとその人柄、考え方に感銘を受けました。

もうひとつは、2012年にロンドンで開催されたパラリンピックです。当時私はイギリスにいて、ロンドンで沸き起こっていたパラリンピックのムーブメントを現地で直に感じる機会に恵まれ、強く興味をもつようになりました。その後、Channel 4が2016年のリオデジャネイロパラリンピックのPR動画として製作した『We're the Superhumans』にも大きな衝撃を受けました。驚くべき実力をもったパラリンピアンが世界中にいる。できるだけ多くの人にその素晴らしさを知ってもらいたい。見てもらいたい。そう願う、つくり手の想いと表現力に圧倒されました。

なのに、2020年に東京でパラリンピックが開催されると決まったとき、素直にはよろこべませんでした。招致のプレゼンテーションの決め台詞の「おもてなし」にも違和感を覚えました。そもそも自分が住む東京という街に、おもてなしの心が本当にあるのか。そんなことを世界中に約束して大丈夫なのか。あるとき国枝さんが、「東京は、車椅子で駅のプラットホームに辿り着くだけでも大変なんです」とおっしゃっていたことも、ずっと心に引っかかっていました。

今の東京は、世界の「多様な人びと」を迎え入れ「おもてなし」できる街とはいえない。そう思うのならば、自分にできることは何だろう。障がいのある方を取り巻く環境や、パラリンピアンの存在を多くの人に知ってもらうための表現や手段を考えるうち、「自分も利用してみたい」と誰もが羨む、障がい者専用施設を都内に作れないだろうか、と発想が切り替わっていきました。

見た人、使った人が驚いて、思わず誰かに話したくなる。そんな障がい者専用施設が東京にできることで、健常者といわれる人たちにも直接的な「気づき」が生まれるのではないか、と。

障がいのある方に話をうかがうと、自分とはまったく違う感覚で日々を送られていることを知り、ハッとさせられることが多々あります。でも、彼らにとっては当然の感覚や判断で、聞けば多くの人がなるほどと思うものも多くあります。そんな「気づき」の場を街のなかに意図的に作り出してみたい。それが、プロジェクトの構想のはじまりでした。

多様性を受け止める、開かれた社会の象徴として

「LGBT」という言葉を聞くようになってから、ずいぶん経ちます。セクシャルマイノリティを表す言葉であり、人の多様性を示す旗印として、浸透したところもあると思います。そのこと自体はとても素晴らしいこと

左：2020年7月完成の「はるのおがわコミュニティパークトイレ」。右：高齢者優先設備などのピクトグラムが並ぶ「恵比寿駅西口公衆トイレ」のユニバーサル・トイレ入口。

Left: Haru-no-Ogawa Community Park Public Toilet completed in July 2020. Right: The universal toilet entrance of Ebisu Station West Exit Public Toilet is lined with pictograms of elderly priority facilities, etc.

だと思っています。と同時に、「LGBT」といったカテゴリーで分けられるほど、人は単純なものではないとも思っています。

多様性について考えていくなかで、次第に、障がいのある方も健常者も、そして、LGBTもQも+も関係ない、すべての人が「違う」という意味で平等な社会、「違うことがあたりまえだと思える社会」のあり方を目指し、表現することが、自分なりの「おもてなし」なのでは、と思うようになっていきました。

みんな違う。だけど、みんなに共通すること。人間が人間である以上、誰しもに関わる根源的な場所。それがトイレだと気づきました。忙しくて食事も睡眠も取れない日はあっても、一度もトイレに行かなかったという話は聞いたことがありません。トイレと無関係でいられる人など、一人もいない。

なのに、街なかの公共トイレは、汚いとか臭いとか、良いイメージがありません。その公共トイレを、驚くほど美しく、誰にとっても使いやすい安全な場所として生まれ変わらせることができれば、多様性を受け止める社会の実現につながる一歩になるのではないか。そう考えるようになりました。

世界から見れば日本のトイレは圧倒的に清潔だと言われています。それでも積極的に使いたい、というものにはまだ遠い。だからこそ、まだまだイノベーションの可能性を秘めた場所だと感じたのです。街なかの公共トイレを、誰もが驚く、新しい気づきの場へと刷新するべく、2018年に「THE TOKYO TOILET」をスタートさせました。

誰もが快適に利用できる公共トイレへ

このような個人的な思いがいくらあっても、共に歩んでくれる人がいなければ、何かを成し遂げることはできません。事業統括を担ってくれた日本財団、全トイレの施工を一括して引き受けてくれた大和ハウス工業、最新設備の提供とトイレ空間自体の監修をしてくれたTOTO、計画に賛同してくれた渋谷区など、数多くの組織や企業の協力を得られたことは本当に幸運でした。何よりも大きかったのは、企画に参加してくださった16組のクリエイターの存在です。

クリエイターの選定理由を聞かれることが多いですが、決めていたのは、公共建築だからといって建築家だけにお願いするのではなく、さまざまな分野のクリエイターにも依頼すること。デザインも多様性が重要だと思ったからです。大切にしたいと思ったのは、各クリエイターとの「議論」です。なぜ新しい公共トイレが必要なのか、これからの公共トイレはどうあるべきかなど、率直な意見交換を行うことは、とても大切なプロセスでした。

すべての施設共通の必須条件として、利用者の性別を問わず、車椅子でも利用でき、オストメイト対応などの機器を備えたユニバーサル・トイレを設置することをお願いしました。それ以外については、敷地それぞれが抱える課題や問題に対して、適切な回答を皆さんが導き出してくれました。安全性の担保、周辺環境への配慮、公共施設が担うべき役割など、皆さんとの対話のなかで、たくさんのこ

とを教えていただきました。

うれしかったのは、施設を作るだけではなく、その後の清掃や維持管理に力を入れたいと考えたことに対して、共感と賛同をいただけたことです。クリエイターのなかには、メンテナンス重視の計画があったからこそ、デザインを引き受けたと言ってくださった方もいました。

海外からの評価や反響の大きさ

供用がはじまって驚いたことのひとつは、海外からの注目度の高さです。特に、2020年夏に完成した透明トイレなどは、SNSでその存在が瞬く間に拡散したこともあり、海外からも予想以上の問い合わせをいただきました。

公共トイレというテーマの潜在能力や可能性を海外で直に感じ、個人的にも大きな経験となったのは、2022年にアメリカのハーバード大学デザイン大学院（GSD）で講演を行ったことと、ミラノサローネでプロジェクトの展示・紹介をしたことです。

ハーバード大学GSDは、建築設計と環境設計、そして都市計画の3学科を軸にしたデザインスクールです。「THE TOKYO TOILET」を建築の観点だけでなく、都市計画や公衆衛生の専門家からも題材として面白い、と思っていただけたことがうれしく、とても貴重なご意見をいただく機会となりました。

ミラノサローネは、年に1度の世界最大規模の家具見本市です。期間中は家具に限らず、さまざまな分野のデザイン展も行われることから、「世界一のデザインの祭典」ともいわれています。そのミラノサローネで「THE TOKYO TOILET」の紹介をすることは、プロジェクト開始当初からの願いでした。なぜなら、このプロジェクトを単に「公共事業の変わり種」ではなく、デザインやアートの文脈で捉えてもらいたい、と考えていたからです。

メイン会場となったのは、地下鉄ドゥオーモ駅内の公共トイレ。森山大道さんが撮影した「THE TOKYO TOILET」の写真で空間を構成し、TOTOのウォシュレットを設置、寄贈するなど、体験を伴う多面的なインスタレーションを行いました。

ありがたいことに反響も大きく、現地で取材をいくつも受けましたが、印象に残っているのは、現地メディアの皆さんが、「トイレをきれいにしてくれて、本当にありがとう」と、感謝の言葉を何度も口にしてくれたこと。そして、「この場所にトイレがあるとは知らなかった」という人が、ほとんどだったことです。

聞けば、公共トイレを使うという選択肢はなく、場所を確かめたことも気に留めたこともなかった、と。「THE TOKYO TOILET」のインスタレーションでその存在に気づき、こんなにきれいなら、困ったときの選択肢になる、と。

課題解決と問題提起を同時に起こす

ミラノサローネでのリアルなその反応は、自分がこのプロジェクトで何がしたかったのか、何を見たかったのかを思い出させてくれる左、右：2022年6月のミラノサローネでの「THE TOKYO TOILET / MILANO」展示風景。ドゥオーモ駅構内のほか、トリエンナーレデザイン美術館でも展示を行った。

Left and right: THE TOKYO TOILET / MILANO exhibition at the Salone del Mobile in June 2020. The exhibition took place at the Duomo station and the Triennale Design Museum.

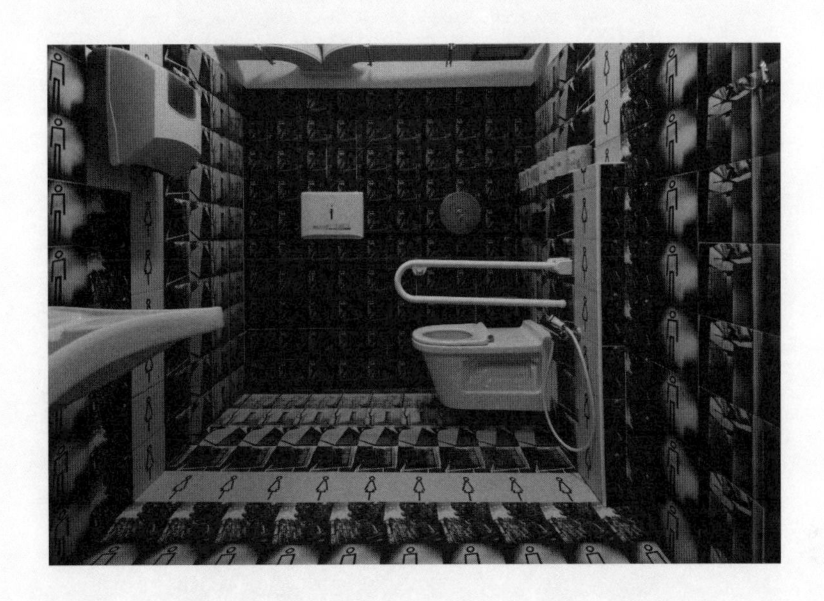

ものでした。今まで気づいていなかったことに気づく、きっかけになること。デザインやクリエイティブの力で、何かが「発見される」瞬間を見ること。もしかしたら、日本より海外の方が、公共トイレを刷新するプロジェクトのインパクトは大きく、その可能性も大きいのではないかと感じたほどです。

「THE TOKYO TOILET」を進めるなかで、何度も思い起こしてきた言葉があります。「デザインは課題解決であり、アートは問題提起である」。

元MIT（マサチューセッツ工科大学）メディアラボの副所長で、グラフィックデザイナーのジョン・マエダさんが語った言葉として、デザインの世界では広く知られています。この言葉に繰り返し立ち返るうち、私は、課題解決と問題提起の両方が同時に達成できたとき、意識や価値観の変容というものが起こるのではないか、と考えるようになりました。

「THE TOKYO TOILET」は、課題解決と問題提起の両方を同時に起こすことを目指しています。汚くて怖い、車椅子利用者や要介護者には使えない――公共トイレが抱えてきた課題を解決すると同時に、課題が存在していることの根本には何があるのかに目を向けることができれば、今より少しは暮らしやすい社会になるのでは、と考え、そのきっかけになりたいと思って立ち上げたプロジェクトです。ミラノサローネで、その意識変革や価値観の変容のカケラ、みたいなものを目の当たりにし、どうやら、このアプローチは間違いではなさそうだ、と確信をより深めることができました。

終わらない挑戦、心を動かすアートの力

一方で、東京・渋谷の現場では、完成後のメンテナンスと向き合う日々が続いていました。2020年の7月から段階的に施設が完成していったのですが、初期に完成したトイレのいくつかは、ひどい汚され方をしたり、外壁に落書きをされたり、修繕が必要となる事態も起きました。

思っていた以上に「公共トイレがきれいに使われ続けること」は難しく、どうすればよいかに苦心し続け、あるとき、クリエイティブディレクターの高崎卓馬さんに相談をもちかけました。

高崎さんからは、「個別のトイレへの対応も大事だけれど、そもそも公共トイレは汚してもいいと思っているマインドセットが問題なのだから、認識をガラリと変えてもらわない限り、事態は好転しないのでは?」、「プロジェクト全体、一連のトイレ群の見え方が劇的に変わる施策が必要なのでは?」とのアドバイスを受けました。「トイレをきれいに使ってください!」と大声で呼びかけるだけではない伝え方、伝わり方。たとえば、芸術的で美しいものに触れたとき、自然と心を動かされるように、アートがもつ「人びとの気持ちを動かす力」を生かしてみるのはどうだろう、と話が進み、「THE TOKYO TOILET」の新たな挑戦として、映画『PERFECT DAYS（パーフェクトデイズ）』が動き出しました。

写真提供: SKWAT（TTT/MILANO パートナー） Photo provided by SKWAT (TTT/MILANO Partner)

左：2023年完成の映画『PERFECT DAYS』ポスタービジュアル。右：同作品より。ロケ地のひとつ「西原一丁目公園トイレ」。主演の役所広司氏は同作品にて第76回カンヌ国際映画祭最優秀男優賞受賞。

Left: Poster image for the film *PERFECT DAYS* completed in 2023. Right: Still photo from the film. It was taken at Nishihara Itchome Park Public Toilet, one of the filming locations. Koji Yakusho, who starred in the film, won the Best Actor Award at the Cannes Film Festival for his role in *PERFECT DAYS*.

監督は、私も高崎さんも昔から尊敬をしていたヴィム・ヴェンダースさん。主役はトイレの清掃員。その主役を演じる俳優は、最初から、役所広司さんにお願いしたいと決めていました。まずは役所さんに打診をし、同時に監督には高崎さんが企画書を、私がお手紙を、自分が考えている課題意識やお願いをしたい理由を精いっぱい書いてお送りし、2021年の年末に、お受けいただける旨のお返事をいただきました。

ヴェンダースさんからのお返事もお手紙でいただいたのですが、まるで映画のワンシーンのような素敵な文面で、受け取ったときの感動は、今でも忘れられません。コロナ禍が続いていた2022年の東京での撮影を経て、映画『PERFECT DAYS』が完成したことは、本当に奇跡のような出来事でした。

自分の想像を超えて、物語が歩き出す

17の「THE TOKYO TOILET」を作る過程では、予想外の障壁が次から次へと立ちはだかり、そのたびにプロジェクトのメンバーや各分野の専門家から斬新なアイデアをいただいて、なんとか乗り越える、その繰り返しでした。

そのすべての道のりにおいて、つくづく感じたのは、自分以外の人の力をお借りし、「自分だけでは思い描けなかったもの」になっていくことが、どれほど重要か、ということでした。施設のデザインはもちろん、映画のような完成後の展開も、自分の発想を超えていくものだからこそ、強く伝わるものになる。多くの人が関わり、多くの視点が加わることで、想像もつかなかった物語が生まれる。今は、その物語がこの社会に生きるすべての人にとって、幸せなものであることを願うばかりです。

早朝や深夜に、完成したすべての施設を見て回るのが好きです。あるときふと、「THE TOKYO TOILET」が、東京の景色の一部になっているな、と感じたことがありました。それはたぶん、参画してくださった一人ひとりのクリエイターが、公共トイレの課題に真摯に向き合って出した答えが、ひとつの提案として受け入れられていることの、証明に近いもののような気がしました。

深い考察のもとに作られた17のトイレは、街に溶け込み、長く愛され、もしかしたら社会を変える気づきとなる。私はそう信じています。

柳井康治 やないこうじ

1977年生まれ。三菱商事勤務を経て2012年、ファーストリテイリング入社。現在、同社取締役、上席執行役員。「THE TOKYO TOILET」は個人の活動として2018年に発案。資金提供やクリエイターの選定から継続的なプロジェクトの企画・ディレクションを行う。また、2023年に映画『PERFECT DAYS』のプロデュースを手がける。

THE TOKYO TOILET:
Raising Awareness and Inspiring Hearts and Minds

Koji Yanai
THE TOKYO TOILET Project Initiator/Sponsor

Creating new public toilets in Tokyo that everyone could use comfortably. I reached this idea by accumulating encounters with various people, personal experiences, and insights from these interactions.

One such encounter was with Shingo Kunieda, the first wheelchair tennis player to achieve a Grand Slam in men's singles. I first had the pleasure of meeting him in 2008. I was deeply impressed by his strength, personality, and way of thinking, which transcended the wheelchair tennis category, even earning praise from the great Roger Federer, who said Kunieda was the best player in Japan's tennis world.

Another one was the London 2012 Paralympic Games. I was in England at the time and had the opportunity to experience firsthand the excitement of the Paralympic movement in London, which sparked my strong interest in the event. Later, I was also greatly intrigued by "We're the Superhumans ," a promotional video for the Rio De Janeiro 2016 Paralympics Games produced by Channel 4. I was overwhelmed by how brilliantly the creators conveyed their desire to let as many people as possible discover and see the amazing Paralympians with incredible abilities all over the world.

Yet I was frankly not thrilled to learn about the decision that Tokyo would host the Paralympic Games in 2020. I also felt uncomfortable with the keyword "Omotenashi" (hospitality) used in the bid presentation. In the first place, I wondered if the city of Tokyo, where I live, indeed has the spirit of Omotenashi. Is it fair to promise such a thing to the world? One day, Mr. Kunieda said, "In Tokyo, simply reaching the train station platform in a wheelchair is a big challenge," a remark that has stuck in my mind ever since.

Tokyo today is not a city that can welcome and offer hospitality to diverse people from around the world. If we think so, what can we do to help? As I thought about the environment surrounding people with disabilities and explored what expressions and means can be used to make the presence of Paralympians known to as many people as possible, I began to ponder the possibility of creating a facility dedicated to people with disabilities that everyone would want to use and admire.

It would be a facility that would amaze those who saw and used it and make them want to share the experience with others. I felt that establishing such a facility dedicated to people with disabilities in Tokyo would raise immediate awareness among the so-called able-bodied persons.

When I talk to people with disabilities, I am often struck by how they approach their daily lives with entirely different senses than mine. However, many of the senses and decisions that are natural to them also make sense to many others once one listens to them. I felt an urge to consciously create such a place of "awareness" in the city. This was the beginning of our project conception.

As a symbol of an inclusive and open society embracing diversity

We have been hearing the word "LGBT" for quite some time now. I think the term referring to sexual minorities has spread as a banner for human diversity to some extent, which is a remarkable accomplishment in itself. At the same time, I also believe that humans are not so simple as to be classified into "LGBT" categories. As I thought about diversity, I gradually realized that perhaps my version of "Omotenashi" is to aim for and express ideas of a society where all people are equal in the sense that they are different and being different is normal, regardless of whether they are disabled, able-bodied, LGBT, Q, or otherwise.

Everyone is different. Yet we all have one thing in common. I realized that the toilet is a fundamental place that concerns all human beings. While we may be too busy to eat or sleep on any given day, I have never heard of anyone who has not gone to the toilet all day. Not a single person can afford to stay away from toilets.

Yet, public toilets in the city evoke negative impressions, such as being dirty or smelly. I began to think that transforming such public toilets into surprisingly beautiful, safe places accessible to all would be a step toward realizing a society embracing diversity. From a global perspective, toilets in Japan are considered by far the cleanest in the world. Still, they are far from what one would prefer to use. That is why I thought toilets still have excellent innovation potential and launched THE TOKYO TOILET project in 2018 to transform public toilets in the city into places of renewed awareness that will surprise everyone.

Public toilets everyone can use comfortably

No matter how much personal commitment one may have, one cannot accomplish anything without partners to walk with. I was indeed fortunate to collaborate with many organizations and companies, such as The Nippon Foundation, which oversaw the project; Daiwa House Industry, which undertook the construction of all the toilets; TOTO, which provided state-of-the-art equipment and supervised the toilet spaces; and Shibuya Ward, which endorsed the project. Above all, I am grateful to the sixteen creator teams who participated in the project.

Many people ask me about our criteria for selecting creators. I had decided beforehand to include creators from various fields rather than only architects simply because they were public facilities. The reason was that I thought diversity was key in design as well. I emphasized the importance of discussion with each creator. Exchanging frank opinions about why new public toilets are needed and what the future of public toilets will look like was a crucial part of the process.

As a standard requirement for all facilities, I asked the creators to install universal toilets that are gender-free, wheelchair-accessible, and equipped with ostomate fixtures. Otherwise, each creator devised appropriate solutions to specific challenges and issues at each site. I have learned a lot through discussions with them, including how to ensure safety and pay attention to the surrounding environment and the role that public facilities should play.

What made me happy was the sympathy and support we received for our decision to focus not only on building the facilities but also on cleaning and maintaining them afterward. Some of the creators told me they agreed to undertake the design because of our maintenance-oriented plan.

High reputation and great response from the world

One of the things that surprised us after the facilities opened for use was the amount of attention it attracted from overseas. In particular, the transparent toilets completed in the summer of 2020 received unexpectedly high numbers of inquiries from overseas, partly due to the immediate publicity surrounding these toilets on social media.

In 2022, I gave a lecture on this project at the Graduate School of Design (GSD), Harvard University in the U.S. I also showcased the project at the Salone del Mobile in Milan, both of which were great experiences for me, allowing me to appreciate firsthand the potential and possibilities of the subject of public toilets overseas. Harvard GSD is a design school comprising three departments: architectural design, environmental design, and urban planning. I was delighted that THE TOKYO TOILET project attracted interest not only from architects but also from urban planning and public health specialists, and it provided an excellent opportunity to receive their valuable opinions.

The Salone del Mobile in Milan is one of the largest annual furniture fairs in the world. It is also known as the world's biggest design festival because it hosts exhibitions in various design fields and furniture during the event. Showcasing THE TOKYO TOILET at the Salone was one of our aspirations from the outset of the project because I wanted people to see this project in the context of design and art rather than simply as an unusual variant of a public facility project.

The main exhibition venue was the public toilets in the Duomo metro station. I presented a multifaceted experience-based installation in the space featuring photographs of THE TOKYO TOILET facilities by Daido Moriyama, which also included the donation and installation of TOTO washlets. Thankfully, I received a great response, and local journalists interviewed me on-site. But one of the most memorable moments was when people from the local media repeatedly expressed gratitude to me, saying, "Thank you for cleaning up the toilets." In addition, many of them also said, "I didn't know there were toilets in this place."

I asked them about this, and they told me they had never considered using public toilets or even bothered to check their locations or paid any attention to them. They became aware of their existence through THE TOKYO TOILET installation. They said they would consider using the toilets when in need because they were so clean.

Solving problems and raising questions simultaneously

The straight-forward responses at the Salone reflecting the reality of public toilets reminded me of what I wanted to do and witness in this project: to help people realize what they had not noticed before and see the moment of "discovering" something through the power of design and creativity. I even thought that the impact of public toilet renewal projects might be more significant overseas than in Japan, and the potential for such a project might be even more profound.

While proceeding with THE TOKYO TOILET project, I have frequently recalled the following phrase: "Design is a solution to a problem. Art is a question to a problem."

This phrase by John Maeda, former associate director of research at MIT (Massachusetts Institute of Technology) Media Lab and a graphic designer, is widely known in the design world. As I kept returning to these words, I began to think that simultaneously solving problems and raising questions may help transform awareness and values.

THE TOKYO TOILET aims to solve problems and raise questions simultaneously. Public toilets are dirty, scary, and inaccessible to wheelchair users and those requiring nursing care. I thought that solving these problems public toilets have faced and simultaneously focusing on the underlying causes of the existing problems would help improve our society into one where one can live more comfortably. This project was launched to provide the impetus for the transformation. At the Salone, I witnessed a glimpse of such a transformation of awareness and values and became more convinced that our approach is not mistaken.

Never-ending challenges and the power of art to move hearts and minds

Meanwhile, maintenance efforts after the completion at the sites in Shibuya, Tokyo, remained a day-to-day struggle. The facilities were completed in stages that started in July 2020. Still, some of the facilities built in the early stages were severely defaced, and graffiti was scrawled on the exterior walls, necessitating repairs.

Maintaining clean use of public toilets was more difficult than I had expected, and we continued to agonize over what to do about it. One day, I approached creative director Takuma Takasaki for advice.

Koji Yanai

Born in 1977. After working for Mitsubishi Corporation, he joined Fast Retailing Co., Ltd. in 2012. He is currently the Director and Group Senior Executive Officer of the company. He conceived THE TOKYO TOILET in 2018 as his personal initiative. He has continuously engaged in the planning and direction of the project, from funding to the selection of creators. In 2023, he produced the film *PERFECT DAYS*.

Mr. Takasaki gave me the following advice: "It is important to deal with individual toilets, but the problem lies in people's mindset to think it is okay to mess up public toilets in the first place. Unless we do something to change this mindset, the situation will not improve... we need to take measures that will drastically change the way people perceive the entire project and the toilet facilities." Instead of loudly calling out, "Please use the toilet more cleanly!" we could perhaps use the power of art to move people's hearts and minds, for example, just like when we naturally feel moved when encountering something artistic and beautiful. This discussion led us to embark on a new challenge for THE TOKYO TOILET: the film *PERFECT DAYS*.

From the beginning, we had decided that we wanted Wim Wenders, whom Mr. Takasaki and I had long admired, to direct the film, and Koji Yakusho to act as the main character, a toilet cleaner. I first approached Mr. Yakusho. At the same time, I wrote a letter to Mr. Wenders and Mr. Takasaki sent it to him with a proposal, writing the best I could about my awareness of the problems and why I wanted to have his involvement. I received his reply at the end of 2021 that he would accept the offer.

His reply was in the form of a letter as well, which was so beautifully written, like a scene from a movie, and I still remember how moved I was upon receiving it. After filming in Tokyo during the continuing COVID-19 pandemic in 2022, *PERFECT DAYS* was completed, which was indeed a miraculous event.

The story transcending our imagination

While constructing the seventeen facilities of THE TOKYO TOILET, we faced a series of unexpected obstacles, which we somehow overcame each time with innovative ideas from the project members and experts in various fields.

Throughout the entire process, we sincerely felt how important it was to enlist the help of others and turn it into something that I could not have envisioned on my own. The facility design and the post-completion development, including the film, transcend my imagination, which is why it conveys such powerful messages. The involvement of many people and viewpoints create a story transcending my imagination. I sincerely hope it will be a happy story for everyone in this society.

I like visiting and looking at all the completed facilities early in the morning or late at night. One day, I felt that THE TOKYO TOILET facilities became a part of the Tokyo landscape. I think this probably means that the solutions presented by each participating creator, who sincerely addressed the challenges public toilets faced, were accepted as valid proposals by the public.

The seventeen toilets, developed through deep considerations, will blend with the city, be cherished for a long time, and possibly become a place of awareness that will change society. I am truly convinced that this will be so.

身近な社会課題を見逃さず、諦めずに取り組む

笹川順平
日本財団 常務理事

日本財団は、社会の課題を解決するためのモデルをつくり、広めていく組織です。障害者の就労や子どもの貧困問題、災害復興など、多岐にわたる社会課題に取り組んできました。社会を良くしていきたい、という思いと、その思いを現実化していくための資金や技術面のサポート、行政との橋渡しなど、これまで培ってきた、さまざまな知恵と経験があります。

その知恵と経験をもってなお、「THE TOKYO TOILET」の運営は難易度が高く、苦労の多いプロジェクトでした。

公共トイレのような「公益性」の高い事業が背負う責任の重さは、計り知れません。公益とは、特定の個人や組織のみではなく、広く社会一般の利益を指します。その「公益」の領域に立ちながら、「個人」の思いと資金を出発点に、クリエイターの意図を汲み、参加企業の理念を理解し、最終的には利用者が「THE TOKYO TOILET」があってよかった、広く社会一般の利益になった、と思えるゴールまで辿り着かなくてはならない。手を取り合って進む人たちの多さや立ち位置の複雑さは、これまでにない、特殊なものでした。

プロジェクトの発案者、柳井康治さんから「公共トイレを刷新したい」というお話をいただいたとき、その視点に驚き、インクルーシブな社会づくりのひとつの切り口として、大きな可能性を感じました。でも、「公共」ですから、個人の思いだけで建替えられるものではありません。そこでまず必要となったのが、公共トイレを管轄する行政との連携です。

当該行政が、なぜ渋谷区になったのかの理由は明確で、日本財団と渋谷区が、「ソーシャルイノベーションに関する包括連携協定」を結んでいたからです。この協定は、自治体と日本財団が双方の強みを生かし、地域の課題解決に連携して取り組んでいこうとするもので、目指していることのひとつには、渋谷ならではの文化や芸術を、世界へ発信することも含まれていました。その協定のもと、渋谷区に打診し、プロジェクトの実践の場を提供してもらうことができました。

プロジェクトチームづくりで最も難航したのは、設計・施工の担い手探しです。プロジェクトがスタートしたのは東京オリンピック・パラリンピック前。建設需要が高まるなか、建物の規模は小さくても手間がかかるうえ、利益の出る仕事ではありません。このような特殊な事業を大和ハウス工業が引き受けてくれたことには、ただただ感謝の一言です。

そして、TOTOのトイレへの専門性の高さなくして、このプロジェクトは成り立ちませんでした。公共トイレに求められるものは何か。TOTOの調査、研究、開発の蓄積とプロジェクトへの共感が、実現を支えてくれました。

クリエイターの方々にとっても、このプロジェクトへの参加は相当な覚悟を有するものだったと思います。だからこそ、携わってくださった方々と利用者の双方にとって、良い結果となることを目指し、両者の間に立って、あらゆる角度から策を練る必要がありました。

建物の完成は道半ば、維持管理が正念場

私たち日本財団が「THE TOKYO TOILET」の運営に力を注いできたのは、先にお伝えしたように、インクルーシブな社会づくりのひとつの切り口として、大きな可能性を感じたからです。もっと言えば、このプロジェクトが、ソーシャル・イノベーションにつながる活動だと信じているからです。

そのイノベーションとは、単に、これまでにないデザインのトイレが誕生する、ということだけを指しているのではありません。公共トイレという「みんなの場所」をみんなで守っていく、社会的な機運を高めること。そのためには、完成したトイレが美しく清潔に使われ続

笹川順平 ささかわじゅんぺい
1975年生まれ。三菱商事、マッキンゼー・アンド・カンパニーを経て、2017年より日本財団常務理事。経営企画および広報を担当し、子どもの貧困対策支援や熊本地震・豪雨災害をはじめとする災害復興支援、ウクライナ避難民支援などを率いる。

け、地域の方々にも歓迎され続けなければなりません。メンテナンスがうまく機能することこそ、このプロジェクトの正念場。「施設の完成だけでは道半ば」なのだと、当初から何度も口にしてきました。

メンテナンスは、日本財団と渋谷区、そして渋谷区観光協会の三者で取り組んできました。清掃の回数を必要に応じて1日3回に増やしたり、第三者機関である「トイレ診断士」による診断を実施したり。その報告書をもとに毎月、維持管理協議会を開き、トイレの利用状況を把握しながら、業務の改善を図ってきました。

「THE TOKYO TOILET」で培った知見は、2024年3月末に維持管理業務が渋谷区に移管したあとも、引き継がれる予定です。問題は資金面で、区民の負担だけに頼らない、最適な方法を探っているところです。

大切なのは、良い面も悪い面も含め、「THE TOKYO TOILET」を取り巻く状況をオープンにしていくことだと思います。良い面でいえば、たとえば、利用調査から多くのトイレで女性ユーザーが増えていることがわかっています。これはとてもうれしい変化です。理由をきちんとヒアリングしていけば、効果のあったことに対して、区の予算をつけることもできるでしょう。悪い面でいえば、落書きや破損などの事例もあること。そういったことも率直に訴え、地域との関係や市民の目を増やしていくことが、今後、「THE TOKYO TOILET」が良い形で使い続けられていくための大切なプロセスだと思います。

もうひとつ、維持管理を支える手がかりになるのは、地域の価値を高める、観光資源としての側面ではないかと考えています。「THE TOKYO TOILET」への海外からの注目度は高く、コロナ禍を経て訪日観光客が増えるにつれ、施設を訪れる観光客も増えています。公共トイレに観光資源保全という視点が加われば、これもまた、新たな予算確保につながります。

「障害のある方々が胸をはって外に出られる社会をつくりたい」というのは私たちのテーマのひとつですが、利用調査では、街なかにユニバーサル・トイレがまだまだ少ないことに気づいた、という声も多くいただきました。ユニバーサル・トイレは機能だけが重視されがちですが、ある車椅子ユーザーのかたが、「THE TOKYO TOILET」体験後に「当事者の視点に立った機能で、それが洗練されたデザインの設備に出合えたとき、自分も街や社会に溶け込んでいる感覚を味わえます」と語ってくださったのも印象的でした。

社会課題への多様なアプローチの可能性

それにしても、今回のプロジェクトに対する反響はとても大きく、公共トイレのあり方に、これほど多くの人が関心を寄せたのは初めてのことではないでしょうか。私たち日本財団としても、苦労が多かった分、「THE TOKYO TOILET」からの気づきや学びは多く、社会課題の解決に対する多様なアプローチやソーシャル・イノベーションに関する手法の幅を、今後も広げていきたいと考えています。

社会のなかで必要とされているもので、課題を残したままのものは、まだまだたくさんあります。このプロジェクトを通じて、ぜひお伝えしたいのは、何事も前向きに「変える」ことができる、ということです。“あたりまえ”を疑い、良いチームと新しい手法で真剣に取り組めば、変えられないものはないと思います。みんながあれだけ諦めていた公共トイレですら、変えることができたのですから。

小さなことでも、身近な「なぜ?」を見逃さず、それを公的な位置づけで考え、行動する。「THE TOKYO TOILET」がそのきっかけになれば、これほどうれしいことはありません。

2022年11月に、広報活動の一環として行った「THE TOKYO TOILETのトイレをめぐるバスツアー」。日本財団による助成事業として、渋谷区観光協会が実施した。メンテナンス関連企業・団体の協力を得て、貸切バスで8つのトイレと清掃の様子を見学。建築やアートに関心のある学生及び社会人を対象にしたツアーで、見学後の交流会では、トイレ診断士との質疑応答なども行い、プロジェクトの趣旨や維持管理の取り組みを紹介した。

The bus tour of THE TOKYO TOILET facilities was held in November 2022 as part of promotional activities. It was subsidized by The Nippon Foundation and conducted by the Shibuya City Tourism Association. With the cooperation of companies and organizations involved in the maintenance, the participants visited eight toilet facilities and observed the cleaning process in a chartered bus. The tour was aimed at students and adults interested in architecture and art. After the tour, an exchange meeting was held, with a Q&A session with a toilet assessor and a presentation of the project's purpose and its maintenance and management efforts.

維持管理活動の一環として、「恵比寿公園トイレ」にて2022年10月に行った子ども向けの清掃体験会。トイレ清掃を通じて公共施設の大切さを知ってもらうことを目的に開催した。このほか「神宮通公園トイレ」でも実施。

A hands-on cleaning workshop for children was held in October 2022 at Ebisu Park Public Toilet as part of the maintenance and management activities. The event was held with the aim of educating children about the importance of public facilities through toilet cleaning. The workshop was also held at the Jingu-Dori Park Public Toilet.

社会課題に関心をもつ高校生2人が、多様性をテーマに、「神宮通公園トイレ」など4つの「THE TOKYO TOILET」を調査。障害に関係なく快適に利用できるかなどを調べ、2020年10月に日本財団ジャーナルで発表。

Two high school students interested in social issues investigated four of THE TOKYO TOILET facilities, including the Jingu-Dori Park Public Toilet, under the theme of diversity. They investigated whether people with and without disabilities can comfortably use these facilities and published their findings in the Nippon Foundation Journal in October 2020.

写真提供：日本財団 Photo provided by The Nippon Foundation

Addressing Familiar Social Issues without Overlooking Them and Never Giving Up

Jumpei Sasakawa
Executive Director, The Nippon Foundation

The Nippon Foundation is an organization that creates and promotes models for solving social issues. We have worked on a wide range of social issues, including job opportunities for people with disabilities, child poverty, and disaster recovery. We aspire to improve society and have cultivated a wealth of wisdom and experience in providing financial and technical support and as an intermediary with government agencies to realize our goals.

With all this wisdom and experience, THE TOKYO TOILET was still a challenging and arduous project to operate.

High-level public interest projects such as public toilets carry immeasurable responsibility. Public interest concerns the interests of society at large, not only specific individuals or organizations. While standing in the realm of "public interest," we had to start with the thoughts and funds of "individuals," consider the creators' intentions, understand each participating company's philosophy, and ultimately reach the goal, namely the realization of THE TOKYO TOILET facilities that will benefit their users and society at large. The sheer number of people we worked with and the complex nature of our position was unprecedented and unique.

When the project initiator, Mr. Koji Yanai, told us he wanted to innovate public toilets, we were surprised at his vision. But we immediately recognized its great potential as a way to create an inclusive society. Given their public nature, however, these facilities could not be reconstructed solely based on personal aspirations. The first thing we needed to do was to establish a partnership with the government authorities in charge of public toilets.

Shibuya Ward was chosen as the project site because The Nippon Foundation and Shibuya Ward had concluded a "Comprehensive Collaboration Agreement on Social Innovation." One of the goals of this agreement included promoting Shibuya's unique culture and arts to the world. Under the agreement, we approached Shibuya Ward, and they provided us with sites to implement the project.

The most challenging part of assembling the project team was finding designers and contractors. The project started shortly before the Tokyo Olympics and Paralympics. With heightened construction demands, it was difficult to find collaborators because the buildings, while small in scale, were labor-intensive and not profitable undertakings. We sincerely thank the Daiwa House Industry Co., Ltd. for undertaking such a unique project.

Also, this project would not have been possible without TOTO's expertise in toilets. TOTO's accumulated experience in research, study, and development and their strong empathy for the project underpinned the realization of what was needed in public toilets.

We believe participating in this project was also a considerable commitment for the creators. That is why it was necessary to stand between the people involved and the users and work out measures from all angles to achieve a favorable outcome for both parties.

Completing buildings is only the halfway point; maintenance and management are crucial

We at The Nippon Foundation have been focusing on the operation of "THE TOKYO TOILET" precisely because, as mentioned earlier, we saw great potential in it as a way to create an inclusive society. More specifically, we believe this project will lead to social innovation.

This innovation does not simply mean the creation of novel toilet designs. It is about building social momentum for people to maintain public toilets, which are a place for everyone. It is about building social momentum for everyone to work together to maintain "a place for everyone" – the public toilets. To this end, the completed toilets must continue to be used neatly, cleanly, and welcomed by the community. A critical point for this project is to ensure

Jumpei Sasakawa
Born in 1975. After working for Mitsubishi Corporation and McKinsey & Company, he has been Executive Director of The Nippon Foundation since 2017. He is in charge of management planning, public relations, and leading the foundation's initiatives, such as child poverty relief and disaster recovery efforts, including the Kumamoto earthquake and torrential rain disasters and Ukrainian refugee support.

its maintenance works well. We have repeatedly stressed from the beginning that completing the facilities is only the halfway point.

Maintaining the facilities has been a three-way effort between The Nippon Foundation, Shibuya Ward, and the Shibuya Ward Tourist Association. The frequency of cleaning has been increased to three times a day as needed, and assessments by a third-party "toilet assessor" have also been conducted. Based on the assessment reports, we have held monthly maintenance and management meetings to monitor the use of the toilet facilities and improve their operations.

We will pass on the knowledge gained from THE TOKYO TOILET project after transferring the maintenance and management operations to Shibuya Ward at the end of March 2024. Financing is one of the challenges, and we are exploring the best way to handle this issue without relying solely on the ward residents for the costs.

We believe the key is openly discussing the situation surrounding THE TOKYO TOILET, including the good and bad aspects. On the positive side, for example, usage surveys indicate that the number of female users is increasing in many toilet facilities. This is a very welcome change. If we can conduct thorough interviews to determine the reasons, the ward may be able to secure a budget for effective measures. On the downside, graffiti and damage have been done to some of the facilities. Frankly discussing such cases, developing relationships with the communities, and attracting the attention of more citizens would be an essential process for the continued beneficial use of THE TOKYO TOILET facilities in the future.

Another critical aspect that could support maintenance and management would be its value as a tourism resource, adding to the value of the area. THE TOKYO TOILET is attracting lots of attention from overseas, and the number of tourists visiting the facilities has been increasing in tandem with the increase in the number of tourists visiting Japan after the COVID-19 pandemic. A new perspective on public toilets in terms of tourism resource preservation would also help to secure new budgets.

One of our themes is to create a society where people with disabilities can go outside openly and confidently. In our usage survey, many people have pointed out that there are still very few universal toilets in the city. Universal toilet facilities tend to focus primarily on functionality. We were deeply touched by the comment of one wheelchair user who, after experiencing "THE TOKYO TOILET," said, "When I encounter a facility that is sophisticatedly designed and functional from the viewpoint of the person concerned, I feel that I am part of the city and its society.

Various possible approaches to social issues

This project has received tremendous attention, and this is probably the first time that so many people have taken such an interest in the state of public toilets. We at The Nippon Foundation have learned a lot from THE TOKYO TOILET, and we will continue to expand our diverse approaches to solving social issues and the range of methods for social innovation.

There are still many issues to tackle in society remaining on the table. Through this project, we hope to convey to the public that one can change everything positively. One can change anything by questioning the norm and working earnestly with a good team and new methods. After all, we successfully changed public toilets, which everyone had given up on.

Never overlook the "Why?" questions about small and familiar things around us. Think and act on them from a public perspective. Nothing would be more gratifying than seeing THE TOKYO TOILET inspire such action.

公共トイレへの態度変容を目指して

長谷部健
渋谷区長

渋谷区は2018年に、トイレの環境整備についての基本的な考え方を示した「渋谷区トイレ環境整備基本方針」を策定しています。目指しているのは、インクルーシブなトイレ環境。人種、性別、年齢、障がいの有無に関係なく、ハードとソフトの両面から「みんな」が使いやすいトイレにしていこうとするものです。また、日本財団と渋谷区は、2017年から5年間の「ソーシャルイノベーションに関する包括連携協定」を結び、社会課題の解決を図る先駆的な取り組みを共に手掛けてきました。

そのような背景のなか、日本財団から「THE TOKYO TOILET」への協力を打診されたとき、渋谷区にとって「こんなにいい話はない」というのがまず最初の、率直な感想でした。「THE TOKYO TOILET」は、公共トイレがもつさまざまな課題の解決を図ろうとする、まさに「ソーシャルイノベーションの先駆的な取り組み」です。そのうえ建設費をはじめとする多くの資金をプロジェクト側が提供してくださるというのだから、断る理由はありません。

参加されるのは世界で活躍するクリエイターの方々。デザインのプロフェッショナルによる新しい視点が入ることで、そこに新しい気づきや発見があるだろうことにも期待を寄せました。これからの公共トイレはどうあるべきか。多種多様な提案が社会へのプレゼンテーションとなり、公共トイレに対する態度変容をおこす、とても意義のある試みだと考えています。

この「態度変容」こそが、私たち行政の切実な願いです。渋谷区を含め、東京都の特別区といわれる23区の公共トイレは、通常、建設費や維持管理費はそれぞれの区の住民税でまかなわれています。つまり、街にある公共トイレは、その街の、地域の人たちのものです。でも、使うのは地域住民「だけ」かというと、そうではない。実はそこに、公共トイレの難しさがあります。

地域のものでありながら、使用者は地域住民に限らない。特に渋谷区は、夜間人口より昼間人口が何倍も多く、区内の公共トイレは、仕事で来た人、食事や買い物に来た人など、この街に「来た」人が使うほうが、圧倒的に多いのです。

すると、どうなるか。残念ながら、雑に使われてしまう。まさにここが「公共性」が問われているところで、「自分のものではない場所」というのは、身勝手なことをしてもいい場所になってしまっている。酷い汚され方や壊され方をすることもあります。本来はむしろ、住民一人ひとりの負担によって支えられている「みんな」の場所こそ、大切にすべきなのに。

それらの汚れや破損の清掃や修繕にも、区民のお金が使われます。住民の方々にしてみれば、住民税は、もっと暮らしが豊かになるための施策に使われた方がうれしいはずです。

公共トイレから少し話が逸れますが、区政が態度変容を目指してきた課題には、街なかのゴミ問題もあります。

たとえば、植え込みにゴミが捨てられていたら、そこはゴミを捨てていい場所だと思われ、ゴミがゴミを呼んでしまう。そこでゴミ箱を設置すると、瞬間的にいっぱいになり、周りまでゴミだらけになる。タバコのポイ捨てによる火災までおこったりする。その対応のために、収集頻度をあげたり、取り締まりの人員を増やしたりすることになり、コストは増える一方なのに、根本的な解決には至りません。だから渋谷区では、公共のゴミ箱を増やすのではなく、ゴミは持ち帰りをお願いしています。自分のゴミは自分で持ち帰るという意識こそが重要で、徐々にではありますが、それがあたりまえだという文化が定着しつつあると感じています。

公共トイレも、汚れが汚れを呼ぶような、前の人の雑な使い方が次の人の雑な使い方を正当化してしまうような負の連鎖を断ち切り

長谷部健 はせべけん
1972年生まれ。博報堂を退職後、NPO法人green birdを設立。2003年4月から渋谷区議会議員を務め、渋谷区男女平等及び多様性を尊重する社会を推進する条例の提案などを行う。2015年より渋谷区長。

たい。それは単にメンテナンスの回数を増やし続けるということではなく、公共トイレはきれいに、大切に使うのが「あたりまえ」となり、常にきれいな状態が保てる社会になることが理想です。使う人自身がよりよい選択を自発的に取れるようになること。「THE TOKYO TOILET」を、ぜひそのきっかけにしたい、と考えています。

注目を集めた先にある、みんなの公共

私は渋谷区の行政に携わる前に、民間企業で仕事をした経験があります。民間の視点で見ると、気づくこと、変えていかなければならないと切に思うことが多々あります。基礎自治体といわれる市町村や特別区の行政というのは、究極のサービス業だと思っています。地域住民と直に接し、サービスを提供する。そう考えると、これまでの公共トイレは、サービスを疎かにしてきた部分が数多くあったように思います。

本当は公共のものほど、多様な人たちに多様な体験価値を提供する質の高いものでなければならない。けれど、公共施設に対しては、どうしても引き算の発想で、とりあえず最低限の用が足せればいいよと、質を重視しない傾向があります。公共トイレの場合は、そもそも「注目されていない」ことも、負の連鎖の元にあったような気がします。誰にも見向きもされず、期待されない場所だから、汚され、壊され、近寄りがたくなる。近寄りがたいからまた、汚され、壊される。

その意味でも、「THE TOKYO TOILET」が注目され、質の高い公共空間を実現した意義はとても大きいと思っています。渋谷区はおかげさまで注目を浴びることが多く、情報発信力も高い街です。スクランブル交差点や明治神宮など、日本のアイコンとなる場所も多く、海外からのお客様がたくさんいらっしゃいます。「THE TOKYO TOILET」の供用当初から、驚くほど多くの海外メディアに注目をしていただいています。コロナ禍を経て、これから観光資源として活用していく動きも増えていくでしょう。日本の建築デザインや施工技術の高さは、世界でもトップクラスです。最初は建築として関心をもった人も、次第にその背景や目的に目を向け、そこからまた社会が変わるきっかけが生まれていくのではないかと思います。

プロジェクトへの協力を決めた当初は、渋谷区内、30カ所以上で展開できればと考えていましたが、土地の権利関係や工事に伴う交通規制などの諸問題から候補地が絞られ、今回の17カ所での実施となりました。地盤の改良や遺跡の調査に時間がかかった場所もありました。もちろん地域の方々への説明にも時間をかけました。プロジェクトの趣旨に耳を傾け、協力してくださった区民のみなさまに心から感謝しています。

これからがスタートです。渋谷区に誕生した17の「THE TOKYO TOILET」を、どう活かしていくか。どんな制度や仕組みを作れば、維持管理も含め、みんなにとって最適な運用となるのか。渋谷の街に住んでいることがうれしい、渋谷の街を訪ねるのが楽しい。そんな街になってゆく、ひとつのきっかけとなるために、しっかり検討し、プロジェクトの趣旨に応えていきたいと思っています。

理想を言えば、地域の皆さんも積極的に維持管理に参加するような仕組みを作りたい。コミュニティで自主運営ができるような公共トイレがあってもいい。民間企業やNPO、近隣の小中学校や大学など、あらゆるステークホルダーが関わる方法もあるはずです。公共トイレがきれいで快適で、安全な場所であることを求めているのは、皆さん同じです。そのための議論を重ねていくことが、今後への大きな一歩だと捉えています。

渋谷区が「THE TOKYO TOILET」の実施場所として提供した、建替え前の該当施設

築30年以上経ち老朽化していた建物や、ユニバーサル・トイレが設置されていない施設もあった。また、2006年に施行されたバリアフリー法などの法整備のもと、更新を進めてきた施設についても、ユニバーサル・トイレへの利用集中など新たな問題が生まれており、区としての「インクルーシブなトイレ環境整備」への課題意識が「THE TOKYO TOILET」参画の背景にある。

はるのおがわコミュニティパークトイレ
Haru-No-Ogawa Community Park Public Toilet

笹塚緑道公衆トイレ
Sasazuka Greenway Public Toilet

恵比寿駅西口公衆トイレ
Ebisu Station, West Exit Public Toilet

恵比寿東公園トイレ
Ebisu East Park Public Toilet

神宮前公衆トイレ
Jingumae Public Toilet

西原一丁目公園トイレ
Nishihara Itchome Park Public Toilet

東三丁目公衆トイレ
Higashi Sanchome Public Toilet

神宮通公園トイレ
Jingu-Dori Park Public Toilet

Public toilet facilities before reconstruction, provided by Shibuya Ward as the sites for THE TOKYO TOILET project.

Some of the buildings were over 30 years old and dilapidated, and some of the facilities were not equipped with universal toilet rooms. In addition, facilities that had been upgraded in accordance with the Barrier-free Law (Law for Promoting Easy Mobility and Accessibility for the Aged and the Disabled) enacted in 2006 and other laws faced a new problem with a high concentration of use of universal toilet rooms. Shibuya Ward's decision to participate in THE TOKYO TOILET project was based on its awareness of the problem concerning the development of a more inclusive toilet environment.

鍋島松濤公園トイレ
Nabeshima Shoto Park Public Toilet

恵比寿公園トイレ
Ebisu Park Public Toilet

代々木八幡公衆トイレ
Yoyogi-Hachiman Public Toilet

裏参道公衆トイレ
Urasando Public Toilet

西参道公衆トイレ
Nishisando Public Toilet

広尾東公園トイレ
Hiroo Higashi Park Public Toilet

七号通り公園トイレ
Nanago Dori Park Public Toilet

幡ヶ谷公衆トイレ
Hatagaya Public Toilet

写真提供：大和ハウス工業、日本財団 Photo provided by Daiwa House Industry, The Nippon Foundation

Aiming to Change Attitudes toward Public Toilets

Ken Hasebe
Mayor of Shibuya Ward

In 2018, the Shibuya Ward established the Shibuya Ward Basic Policy for Toilet Environment Improvement which outlines a fundamental approach to improving toilet environments. The goal is to create inclusive toilet environments that are easy to use for everyone, regardless of race, gender, age, or impairment, in tangible and intangible aspects. Meanwhile, The Nippon Foundation and Shibuya Ward signed a five-year Comprehensive Collaborative Agreement on Social Innovation in 2017 and worked together to undertake pioneering initiatives to solve social issues.

Against this background, when The Nippon Foundation approached us about participating in "THE TOKYO TOILET," our immediate and frank impression was that this was going to be a great opportunity for Shibuya Ward. THE TOKYO TOILET is a pioneering social innovation initiative that addresses various issues public toilets face. Moreover, the project team offered to provide much of the funding, including the construction costs, and we had no reason to decline.

Those participating in the project were internationally acclaimed creators. We had high expectations that the fresh perspectives of design professionals would bring new insights and discoveries to the project. How should the public toilets of tomorrow evolve? This project is a significant initiative in which diverse proposals presented to society will trigger changes in attitudes toward public toilets.

This "attitude change" is what we at the administration earnestly hope for. Construction and maintenance costs of public toilets in the 23 special wards of Tokyo, including Shibuya Ward, are usually financed by resident taxes of the respective wards. In other words, the city's public toilets belong to the local residents and community. But they are not used only by the local residents. This is precisely where the difficulty of public toilets lies.

While the facilities belong to the community, the users are not limited to the local residents. Given that the daytime population in Shibuya Ward is several times greater than the nighttime population, public toilets in the ward are predominantly used by people who come from out of town, whether for work, dining, shopping, or other reasons.

Then, what happens? Unfortunately, they get used in an unhygienic manner. Here, "publicness" is at stake. Places one does not own have become places to do whatever one wants. Sometimes they are severely defaced or damaged, even though they are places for all to use, supported at the expense of all residents, and should be cherished.

The ward residents' tax money is also spent on cleaning and repairing these damages. The residents would be happier if their tax money would be used for measures to improve their lives.

To digress slightly from public toilets, the ward government has sought to promote "attitude change" toward the city's garbage.

For example, if someone dumps garbage in a planting, people assume it is acceptable to litter there, and it will attract more garbage. If a garbage can is installed there, it will instantly fill up, and the area around it will overflow with garbage. Cigarette littering can even cause fires. As a result, the frequency of collection and the number of enforcement personnel must be increased, which only adds more costs, but does not lead to a fundamental solution to the problem.

For this reason, Shibuya Ward requests that people take their garbage home with them instead of increasing the number of public garbage cans. The awareness of bringing one's garbage back home is crucial, and I feel that this practice is gradually taking root in our culture as the norm.

Public toilets also need to break the negative cycle in which dirt attracts dirt, and the unclean use of the previous person justifies the sloppy use of the next person. It does not simply mean that the frequency of maintenance should be increased. Ideally, our society should evolve into one in which using public toilets cleanly and with care is common sense, so they are always kept clean. We hope that THE TOKYO TOILET facilities will encourage

Ken Hasebe

Born in 1972. Founded the non-profit organization "green bird" after retiring from Hakuhodo. He served as a member of the Shibuya Ward Assembly starting in April 2003, where he proposed an ordinance to promote a society respecting gender equality and diversity in Shibuya Ward. Since 2015, he serves as the Mayor of Shibuya Ward.

people to make better choices on their initiative.

Public facilities for all, beyond the immediate attention

I have worked in the private sector before joining the Shibuya Ward administration. From a private-sector perspective, I have noticed many things which I sincerely believe need to be changed. I believe the administration of municipalities and special wards, known as basic local governments, is the ultimate service enterprise. They are in direct contact with residents and offer services to them. In this regard, I believe that public toilets have neglected many service aspects in the past.

As far as spending money is concerned, the marketing principle is to clarify what kind of service the money spent is converted into and whose satisfaction it contributes to. However, to date, public toilets have not brought happiness or excellent service to ward residents to a large extent, despite the money being spent by the ward residents. This doesn't seem right.

The truth is that the more public something is, the higher quality it should be, offering diverse experience values to diverse people. However, when it comes to public facilities, people tend to think with a subtractive mindset, and not focus on quality, saying that meeting the minimum needs is good enough for the time being. As for the public toilets, the lack of attention seems to be the source of the negative chain of events in the first place. It is a place no one looks at, or expects much from, so it gets defaced, damaged, and becomes more unapproachable. Because it is unapproachable, it gets defaced and damaged even more.

In this sense, THE TOKYO TOILET has attracted much attention, and the significance of realizing such a high-quality public space is profound. Shibuya Ward is a city with solid information dissemination capabilities, drawing significant attention. It boasts many iconic Japanese landmarks, such as the Scramble Crossing and Meiji Jingu Shrine, attracting visitors from abroad.
THE TOKYO TOILET facilities have received surprisingly large number of attention from the international media since its opening. After the COVID-19 pandemic, we expect a growing trend to use the facilities as tourism resources. Japan's architectural design and construction techniques are among the best in the world. Those initially interested in architecture will gradually focus on its context and purpose, and I hope this will trigger further social changes.

When we decided to participate in the project, we had hoped to expand it to more than 30 locations in Shibuya Ward. However, various issues, such as land ownership and construction-related traffic restrictions, narrowed down the number of potential sites to seventeen. Some places required time to improve the ground and conduct archaeological surveys. We also spent a lot of time explaining the project to the residents. Our deepest gratitude goes to the ward residents who were willing to listen and understand the purpose of the project and cooperate with us.

This is just the beginning. What we need to think about from hereon is, how can we make the most of the 17 public toilets that THE TOKYO TOILET has created in Shibuya Ward? What kind of system and structure should be established to optimize the operation for everyone, including maintenance and management? We want to make Shibuya a city where people feel happy to live and enjoy visiting. We should carefully examine these aspects to meet the aim of the project.

Ideally, we would like to establish a system in which residents actively participate in the maintenance and management of these facilities. There could be public toilets that the community can voluntarily operate. Perhaps there are ways to involve all stakeholders, including private companies, NPOs, nearby elementary and junior high schools, universities, etc. Everyone wants public toilets to be clean, comfortable, and safe. We see the repeated discussions to achieve this as a significant step toward the future.

優しさを象徴する空間としての公共トイレ

吉田伸典
TOTO 執行役員 特販本部長

「THE TOKYO TOILET」にお声がけいただいたのは、2018年の7月のこと。2020年に予定されていた東京オリンピック・パラリンピックに向け、日本が世界に誇るべき文化は何か、多様な人びとを迎えるためにできることは何かを考え、そこで公共トイレに白羽の矢が立ったというお話だったと記憶しています。TOTOは、水まわりの製品を幅広く手掛けていますが、とりわけ衛生設備機器のリーディング・カンパニーとして認知していただいており、お声がけいただいた際も「トイレといえばTOTOですから」と言っていただき、光栄でした。

プロジェクトに参画した理由は、TOTOの衛生設備機器の技術やトイレ空間を紹介し「日本を世界のショールームに」するまたとない機会であること、「世界中にTOTOファンを増やしたい」という想いからでした。個人的には、とにかく、「トイレという空間に着眼していただけたこと」がうれしかった。多様性を受け入れ、誰もが暮らしやすい社会を目指すというプロジェクトの思想にも強く共感し、体現する場所として、トイレ空間が選ばれたことがうれしく、どんどんのめり込んでいったというのが、気持ちとしての正直なところです。

多様性を受け入れよう、誰もが暮らしやすい社会を目指そう、というのはつまり、もっと人に優しくなろう、ということではないでしょうか。とかく自分本位になりがちな時代のなかで、他者を思いやり、優しさをもって物事にあたること。その意識変革がソーシャル・イノベーションであり、公共トイレは、次に使う人のためにきれいに使う、といった他者への思いやりが反映される究極の場所でもあります。今の社会に必要な「優しさ」を象徴する場所なのだと、今回、改めて思うことができました。そのような視点で、名だたるクリエイターとともに17ものトイレを刷新していくというプロジェクトはこれまでになく、その試みに携われたことに、心から感謝しています。

培ってきた技術と経験を活かして

「THE TOKYO TOILET」において、TOTOはプロジェクトのアドバイザーとして、現状調査と設置機器を含めたトイレ空間のレイアウト提案を行いました。TOTOが国内初の腰掛式水洗便器の開発に成功してから、来年で110年になります。公共トイレに関しても、当社は半世紀以上にわたり、調査、研究、開発を重ねてきた歴史があります。また、渋谷区代々木には、設計や施工のプロに向けた専門の提案型ショールーム「TOTOテクニカルセンター東京」を有しています。ここでは、建築家やデザイナーの設計事案にも数多く携わってきた経験がありますので、今回はそのテクニカルセンターのスタッフにも提案作りに加わってもらいました。

トイレは、条件や制約の多い特殊な空間です。トイレとしてきちんと機能するために必要な寸法があり、そこに収める機器にも相応の寸法があります。「THE TOKYO TOILET」では、どの場所にもユニバーサル・トイレを設けたため、必要な機器が多く、おのずと条件や制約も多くなりました。特に、これまでユニバーサル・トイレを設けていなかった敷地面積が限られた場所では、それらを効率良く収めるための個室のレイアウトなど、検討事項が数多くありました。

トイレを初めてデザインするクリエイターには、戸惑いもあったと思います。それぞれの建築コンセプトやデザインに寄り添いながら、施工性やメンテナンス性までを考え、快適なトイレ空間を実現する。そこに、私たちのこれまでの経験や情報の蓄積が強みとして活かされ、アドバイザーとしての役割を果たせたのではないかと思います。

使用する機器の選定で指針としたことのひとつは、環境に配慮した製品であること。具体的には、節水と節電です。今回は手洗い場の水栓のほとんどを、非接触の自動水栓にしています。自動水

吉田伸典 よしだしんすけ

1962年生まれ。1985年に東陶機器（現TOTO）入社。販売統括本部商品営業推進部長、特販本部副本部長などを経て、2017年より現職。「THE TOKYO TOILET」では、社内プロジェクトチームの陣頭指揮を執り、販売・製造・研究開発・プロモーション部門をまとめ上げ、プロジェクトを推進した。

栓は衛生面に注目されることが多いのですが、水の止め忘れがなく、節水につながります。また便器洗浄やウォシュレットの操作には、「エコリモコン」を採用しています。ボタンが押されるたびに発電・充電するため、電池が不要です。

公共トイレですから、メンテナンス性も重要です。先の「エコリモコン」は電池交換が不要で、メンテナンスの軽減という意味でもメリットがあります。また、大便器も小便器も壁掛け式で床から浮いているタイプを提案の軸とするなど、掃除のしやすさには特に気を配りました。17の施設全体を通して、ある程度製品を統一しているのは、後々、破損などで部品の交換が必要となった場合の、対応のしやすさを考えてのこと。その点は、維持管理を担う日本財団や渋谷区のご意向も含め、提案内容を固めていきました。

挑戦が実を結ぶときは必ず来る

トイレ掃除に携わる方々の多大な労力と人員不足も、広く知っていただきたい大きな課題です。

TOTOでは、かねてより、便器表面の汚れが付きにくく、付いても取りやすくする焼成加工技術を開発するなど、メンテナンス性の向上に力を注いできました。今回参画したことにより、より効率よく維持管理ができるシステム開発に取り組む必要性も感じています。企業としてIoTを使った製品を手がけはじめていますが、ブラッシュアップさせることにより、公共トイレの清掃・管理効率を好転させることもできるかもしれない。インターネット接続や相互制御のようなシステムを、多くの人びとが頻繁に利用する公共トイレで使うまでには、いくつものハードルがありますが、取り組む価値のある課題だと考えています。

プロジェクトに参画して5年あまり。完成した「THE TOKYO TOILET」を見て、TOTOの技術や製品を知っていただけるのは、とてもうれしいことです。でも、それより喜ばしいのは、建築デザインを含め、TOTOだけではなく多くの企業や個人が関わって実現した空間そのものを評価していただけることです。もっと言えば、先に申し上げたプロジェクトの思想が利用者に伝わること。それが一番の願いです。

公共トイレは、これから変わると思います。多くの人びとが利用する、広い意味での「公衆トイレ」は、すでに変わりはじめていると感じています。例えば駅構内のトイレは以前に比べ、とてもきれいに使われるようになりました。鉄道会社がトイレの改修に取り掛かり、清潔を維持する意識の醸成を狙い、有料（チップ制）のトイレを採り入れるなど、さまざまな試みが行われてきました。高速道路のサービスエリアのトイレにおいても15年以上前からトイレ整備に力を入れ、今やそれぞれのエリアで快適性を競うほどになっています。昔は汚れや臭い、落書きなどがあり、今日の快適空間とは大きく異なるため、若い世代の人には想像もつかないかもしれません。

そのような状況に変わるまでには、長い年月がかかりましたし、まだ足りない部分はあります。場所ごとの個性を活かす試みは、はじまったばかりです。そこにきて、「THE TOKYO TOILET」は、課題山積のまま残されていた公共トイレを対象に、まさかの、全敷地オリジナルデザイン、全室ウォシュレット装備ですから、大きなチャレンジだと思います。

でも、そのチャレンジが実を結ぶときは必ず来ます。誰もが他者を思いやり、人に優しくなる。そんな社会の実現を、このプロジェクトに関わったすべての方々をはじめ、利用者の皆さまと一緒に見ることができる日を、楽しみにしています。

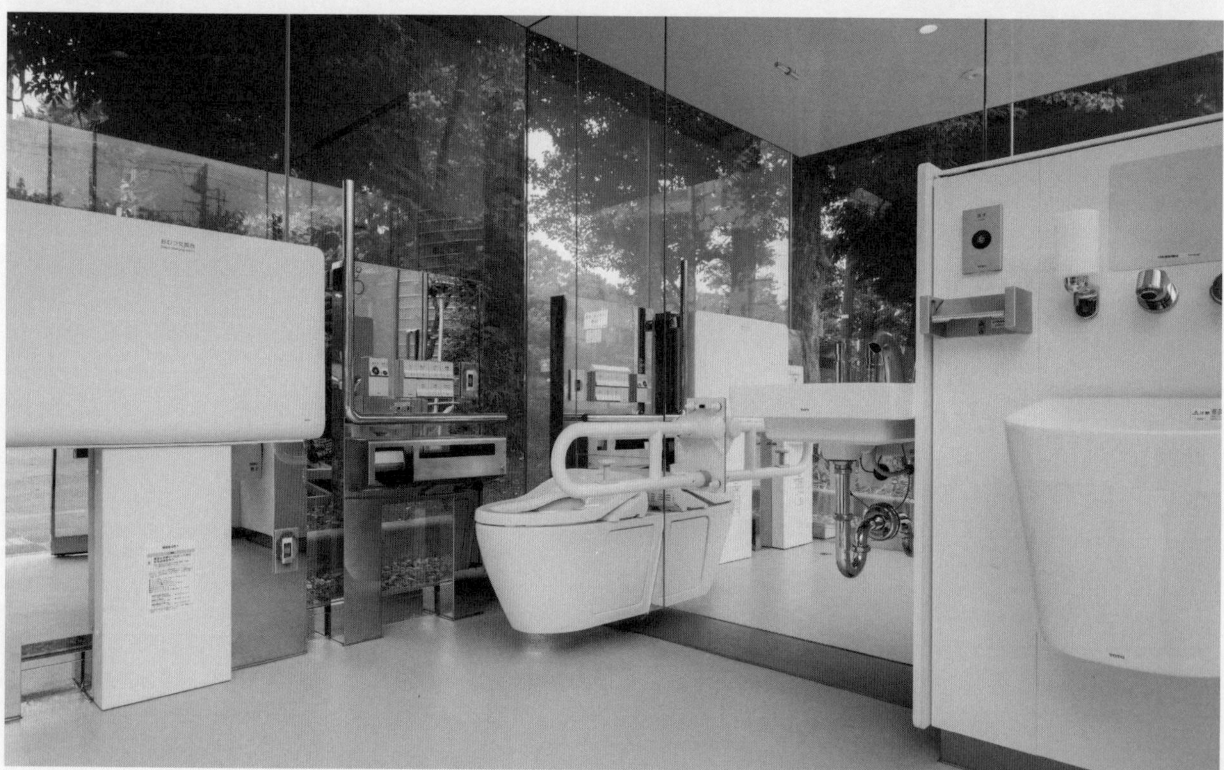

「THE TOKYO TOILET」では、全17施設すべてにユニバーサル・トイレを設置。上：「神宮通公園トイレ」（P.72）では、車椅子使用者やオストメイト対応器具がセットされた〈コンパクト多機能トイレパック〉を採用。下：「はるのおがわコミュニティパークトイレ」（P.22）。〈コンパクトオストメイトパック〉を設置。上下ともに、大便器は〈壁掛大便器セット・フラッシュタンク式〉で、〈ウォシュレット アプリコットP、温風乾燥付きエコリモコン〉。乳幼児連れに配慮したベビーシート、ベビーチェアも備え、さまざまな利用者に対応できる設備を完備した。

All seventeen facilities in THE TOKYO TOILET project offer universal toilet rooms. Top: Jingu-Dori Park Public Toilet (p.72) uses a wheelchair-friendly and ostomate-friendly toilet unit. Bottom: Haru-no-Ogawa Community Park Public Toilet (p.22) is equipped with an ostomate sink. Both facilities in the top and bottom photos feature wall-hung toilets with flash tanks and Washlets with Eco Remote Controllers and warm air dryers." They are also equipped with baby changing stations and baby seats to cater to diverse users.

左上：「東三丁目公衆トイレ」（P.42）の男性用トイレ内。手洗い場の水栓は、非接触で吐水・止水操作ができる〈台付自動水栓〉。右上：「代々木八幡公衆トイレ」（P.112）の〈自動洗浄小便器〉。プロジェクト全体を通して、小便器も床の清掃性に優れた壁掛式とした。右下：「鍋島松濤公園トイレ」（P.92）の子ども用トイレ内。グリップ付きの〈幼児用小便器〉と〈壁掛手洗器〉。左下：「笹塚緑道公衆トイレ」（P.162）の子ども用トイレ内。〈幼児用大便器〉の座面高は275mm。

Top left: Interior view of the men's toilet room in the Higashi Sanchome Public Toilet (p.42). The faucet in the hand wash basin is an "automatic faucet with a pedestal," allowing non-contact water dispensing and shut-off operations. Top right: Automatic Flushing Urinals in the Yoyogi-Hachiman Public Toilet (p.112). Throughout the project, wall-hung urinals were used for easy floor cleaning. Bottom right: Interior view of the children's toilet room in the Nabeshima Shoto Park Public Toilet (p.92). A toddler urinal with grips and a wall-hung type handwash basin. Bottom left: Interior view of the children's toilet room in the Sasazuka Greenway Public Toilet (p.162) featuring an infant toilet with a seat height of 275mm.

Public Toilets as Spaces Symbolizing Kindness

Shinsuke Yoshida
Executive Officer and General Manager of Project Sales Division, TOTO LTD.

TOTO LTD. was invited to participate in THE TOKYO TOILET project in June 2018. At the time, we were thinking about what kind of culture Japan should proudly present to the world and how to welcome diverse people to the Tokyo Olympics and Paralympics scheduled for 2020, and public toilets were chosen as one of the key focus areas. TOTO manufactures a wide range of water-related products and is recognized as a leading company in sanitary fixtures. When those involved in the project approached us, we were honored to be told that "When it comes to toilets, TOTO is the answer."

We decided to participate in the project because it was a once-in-a-lifetime opportunity to introduce TOTO's sanitary fixture technology and toilet space and we wanted to make Japan a "showroom of the world" and increase the number of TOTO fans around the world. I was delighted to see how the project focused on the toilet space. I also strongly sympathized with the project's philosophy of embracing diversity and striving for a society where everyone can live comfortably. I was delighted they chose the toilet space to embody this philosophy, and I became increasingly immersed in the project.

Embracing diversity and striving for a society where everyone can live comfortably means being kinder to others. Today, we tend to be self-centered, but we must remember to be considerate to others and approach things with kindness. This change in awareness is social innovation, and public toilets are the ultimate place to demonstrate our kindness and consideration for others, for example, by keeping them clean for the next user. This project reminded us once again that toilet facilities are a place symbolizing the " kindness" needed in today's society. The project to renew seventeen toilets from such a perspective, with renowned creators, is the first of its kind, and we are truly grateful to be part of this endeavor.

Leveraging the skills and experience we have cultivated

TOTO served as a project advisor for THE TOKYO TOILET project, surveying the current situation and proposing toilet space layouts, including the equipment installation. Next year will mark 110 years since TOTO successfully developed Japan's first western-style flush toilet. With regard to public toilets, we have a history of study, research and development spanning more than a half century. We also have a specialized proposal-based showroom, TOTO Technical Center Tokyo, in Yoyogi, Shibuya-ku, Tokyo, for design and construction professionals. The Technical Center has experience working with architects and designers on many design projects, and we asked the staff there to join us in the proposal development.

Toilet facilities are special spaces with many conditions and constraints. There are dimensions required for a toilet to function adequately, and the devices that come with it also have appropriate dimensions. Since all facilities in THE TOKYO TOILET project included universal toilet rooms, a number of requirements and restrictions increased due to the many devices needed. In particular, in locations with limited site areas where the facilities did not have universal toilet rooms before the renewal, there were many issues to consider, such as the layout of cubicles to accommodate these devices efficiently.

We believe creators designing toilets for the first time must have been at a loss. While staying true to their respective architectural concepts and designs, they also had to consider factors such as ease of installation and maintenance to create a comfortable toilet space. To this end, we utilized our accumulated experience and information to our advantage and fulfilled our role as an advisor.

One of the guiding principles in selecting equipment to use was that the products should be environmentally friendly, specifically regarding water and power conservation. In this project, most faucets at the hand wash basins are non-contact and automatic. Automatic

Shinsuke Yoshida
Born in 1962. Joined TOTO KIKI LTD. (currently TOTO LTD.) in 1985. After serving as Manager of Product Sales Promotion Department, Sales Management Division, and Deputy General Manager of Project Sales Division, he has been serving his current position since 2017. In the "THE TOKYO TOILET" project, He spearheaded the in-house team and promoted the project by organizing the sales, manufacturing, R&D, and promotion divisions.

faucets are often associated with hygiene, but they also save water by never forgetting to turn off the water. "Eco Remote Controllers" are used for toilet flushing and Washlet operation, which generate and recharge whenever a user presses a button, eliminating the need for batteries.

Since these are public toilets, ease of maintenance is also important. The Eco Remote Controller mentioned above does not require battery replacement, offering an advantage in reducing maintenance. We also paid special attention to ease of cleaning, for example, by proposing wall-hung type toilets and urinals raised off the floor. The products used are standardized to some extent throughout the seventeen facilities for easy handling in case components need replacement later due to damage or other reasons. In this regard, we finalized our proposals, taking into account the opinions of The Nippon Foundation and Shibuya Ward in charge of their maintenance and management.

Challenges will always bear fruit

The tremendous labor and lack of personnel to clean public toilets is another significant issue that needs wider recognition.

TOTO has long focused on improving the ease of maintenance by, for example, developing a baking process technology that makes toilet surfaces stain-resistant and easy to clean. Our participation in this project has made us aware of the need to establish a more efficient maintenance management system. For this reason, our company is beginning to develop products based on the IoT, and further improvement of these products could potentially improve the efficiency of cleaning and management of public toilets. There are a number of hurdles to overcome before using systems incorporating the Internet connection and interactive control features in public toilets frequented by many users, but we believe it is a challenge worth tackling.

It has been over five years since TOTO participated in the project. It is our great pleasure to see the completed facilities of THE TOKYO TOILET and to introduce TOTO's technologies and products to the public. But what is more rewarding is the recognition of the space itself, including the architectural design, which was realized not only by TOTO but also by the involvement of many companies and individuals. More importantly, our greatest hope is to convey to the users the project's philosophy mentioned earlier.

We believe that public toilets are going to change from now on. In the broad sense of the word, "public toilets," which many people use, are already changing. For example, toilet rooms in train stations are now much cleaner than they used to be. Railroad companies initiated various efforts to renovate their toilet rooms, including adopting pay-per-use (tipping) toilets, to foster an awareness of maintaining cleanliness. For more than 15 years, highway rest areas have also put a lot of effort into toilet maintenance, and now they even compete with each other over the comfort level of the toilet rooms. In the past, there were stains, odors, and graffiti, which may be unimaginable to the younger generations, as they were very different from today's comfortable spaces.

It has taken many years to change to present-day conditions, and some areas still need further improvement. The efforts to capitalize on the unique characteristics of each location have only just begun. In this context, THE TOKYO TOILET project was a great challenge because it targeted public toilets, an untapped area with a mountain of unresolved issues, and the project surprisingly features new designs for all locations and are equipped with Washlet devices in all rooms.

But the time will surely come when the challenges will bear fruit. Everyone will be considerate and kind to others. We look forward to witnessing the realization of such a society someday together with all those involved in this project and all the users.

建築の限りない可能性に触れて

竹林桂太朗
大和ライフネクスト 代表取締役社長

大和ハウス工業が設計・施工する建物は、社内の技術者だけで完結することがほとんどです。コストや施工期間を守り、クライアントにご満足いただける建築物を完成させ、引き渡すまでが技術者の仕事です。もともと鋼管構造建築からスタートし、「建築の工業化」を理念に掲げている会社ですから、まったく予想外の着地点を求めることはありません。

長年培ってきたそのような自分たちのスタイルと、「THE TOKYO TOILET」の設計・施工は、まったく肌合いの異なるものでした。建築界のノーベル賞といわれるプリツカー賞を受賞された建築家から、ファッションや広告の世界で活躍するクリエイティブ・ディレクターまで。16組のクリエイターが発想する公共トイレは、それぞれ個性的で、施工についても一筋縄ではいかないだろう——当初から、そのように想像するのは容易なことでした。

しかも、お声がけいただいた当時は、東京オリンピック・パラリンピックを控え、金融緩和などの経済政策により景気が上向いているなか、ゼネコン各社の忙しさはすでに目いっぱい。人繰りなどの問題から、お受けするのはどの会社も難しかったと思います。当初は私たちも、お断りをする予定でした。

それでもお受けすることにしたのは、このプロジェクトがトイレという「誰もが必要とする場所」を扱うものであること、「誰もが使えるという公共性」に根ざしたものであるからに他なりません。根底にあるのは「儲かるからではなく、世の中の役に立つからやる」という大和ハウス工業の創業者、石橋信夫の言葉です。その創業者精神に通じるプロジェクトだと思ったから、覚悟を決め、取り組むことにしたのです。

第一線で活躍するクリエイターとやりとりし、利用者によろこんでいただける公共施設をつくるプロジェクトに参画することができれば、その経験は、プロジェクトメンバーのみならず、全社的にも、得難いものになるだろうという確信もありました。特に、これまで社外の方々との共同事業をしたことのない若手にとっては、千載一遇、まさに願ってもない機会です。

私たちは通常、建物の用途ごとに分かれた事業部制で業務を行っています。「THE TOKYO TOILET」では、その事業部の垣根を越え、建築事業部と流通店舗事業部、そして集合住宅事業部の3つの部署を横断するプロジェクトチームを組みました。

参加社員は立候補で募りました。手を挙げる社員は多く、20代から30代の若手を中心に、なかには役職を持つベテラン社員もいました。半数以上を女性が占め、女性の現場監督が多かったことも、今回のプロジェクトチームの特長のひとつです。

話が先に飛びますが、最も心を砕いたのは、施工管理の、管理の部分かもしれません。公共トイレは、一般的な建設現場と比べて敷地条件が複雑な場合が多く、駅前のような人通りの多い場所は工事時間も制限されますし、公園内には車両を入れられないので、資材搬入が難しくなります。遊具や砂場が近い場合は、工事時間外に子どもが囲いのなかに入ってしまう心配もあります。やはり、事故だけは何としてでも防がなければならない。どの敷地にもそれぞれ懸念要素があり、渋谷区や東京都、警察とも事前協議を重ね、安全に作業できるよう慎重に計画を立てました。

難局にも粘り強く、デザインの力を信じて

個人的に「THE TOKYO TOILET」の設計・施工における最初の試練といいますか、腹が据わったのは、槇文彦氏の「恵比寿東公園トイレ」だったように思います。いくつもの曲線が組み合わされた鉄板の屋根。あの屋根は、宮城県・気仙沼にある工場で造船技術を用いて製作し、現場で溶接しています。私たち

竹林桂太朗 たけばやしけいたろう
1969年生まれ。大和ハウス工業の東京本店建築事業部長、東京本店長などを経て、19年に同社の執行役員に就き、建築事業本部営業統括部長などを歴任。2022年4月より大和ライフネクスト代表取締役社長。

は造船に携わる方々と仕事をしたことがありませんでしたし、最初は予算的にも工程的にも、どうしたものかと悩みました。でも、必要なら、やるしかない。

田村奈穂氏の「東三丁目公衆トイレ」は、敷地だけ見たら、本当にここですか？ と目を疑うくらい狭い。ところが出来上がってみると、ユニバーサル・トイレをはじめ、3つのトイレがきちんと収まり、とても快適な空間になっています。そのデザインの力を目の当たりにすると、予算であれ時間であれ、物理的な問題であっても、私たちに無理だとか出来ないという言葉の選択肢はないな、と。そうして5年あまりのあいだ、「これまでにない公共トイレ」を実現するというプロジェクトの趣旨に何度も立ち返り、クリエイターのプランを受け止めて具現化する努力を重ね続けました。

いくつもの挑戦が、かけがえのない財産に

17の「THE TOKYO TOILET」が無事完成を迎え、プロジェクトチームのメンバーに、改めて担当したトイレの見どころを聞いたところ、それぞれが熱い想いを語っていました。物流施設のような大型建築物では、工程や部分ごとに担当が割り振られるのが常ですが、公共トイレは規模が小さいゆえ、施設全体を最初から最後まで見届けることができ、思い入れも強いのでしょう。何より、自ら手を挙げて参加したプロジェクトですから。終始、メンバーのモチベーションは高く、そのモチベーションの高さが、今回の私たちの一番の強みだったかもしれません。

「THE TOKYO TOILET」には鉄筋コンクリート造の建物がいくつかありますが、外観もコンクリート打ち放しの場合、鉄筋数を増やしたり、コンクリートの材質を変えたり、新しい型枠を用意するなど、可能な限り美しく仕上がるよう気を配りました。なかには、四隅を曲面とするコンクリート打ち放しもあり、型枠の割付などに並々ならぬ工夫を重ね、その三次元曲面を実現しています。

塗装にも物件ごとに担当者のこだわりがあります。鉄板には遮熱塗装を、白の外壁には、清掃の負担を減らすべく、太陽光で汚れを分解する光触媒塗装を施しています。

マーク・ニューソン氏の「裏参道公衆トイレ」の銅葺き屋根も、記憶に残る施工事例のひとつです。社寺建築を手がける職方に依頼したのですが、マーク氏がイメージする銅葺き屋根は、日本の寺社仏閣の基準や考え方とはまったく異なるもので、専門の職方にとっても、新たな挑戦となりました。銅板のサイズもイレギュラーで、下地の木も鉄骨の骨組みも3Dを駆使して検討。丸みを再現するため、現場での加工や削り出しにもこだわりました。

現場が後半になると、担当者が納得のいくまで挑み続けるなど、頼もしく感じることも多々ありました。その粘り強さや完成度の高さを褒めていただける結果となり、クリエイターが思い描いたものを体現できたのではないかと自負しています。

このプロジェクトに参画することで、想像を超えるようないくつものハードル、しかもかなりの高さのあるハードルを跳ばなければならないこともたびたびありましたが、実現に至るまでの折衝、調整、設計・施工は、大和ハウス工業のかけがえのない財産になりました。今の願いは、完成した「THE TOKYO TOILET」が大切に使われ、社会の役に立つこと。ひとつとして同じもののない公共トイレと、建築の限りない可能性を、ぜひ現地で実際にご利用いただき、体験してくだされば幸いです。

神宮通公園トイレ **Jingu-Dori Park Public Toilet** [P.72]

建替え前は明治通りを背にした長方形の施設だったが、安藤忠雄による円形平面の建築計画にあわせて敷地を再整備。個室を収める鉄筋コンクリート造のコアを立ち上げ、楕円の屋根やアルミパネルの曲面壁を設置していった。この曲面壁の周囲に、斜めの下がりのルーバーをめぐらせ、通路とした。現場では、壁やルーバーの収まりにもこだわった。

Before the reconstruction, the facility was rectangular, with its back against Meiji Dori. The site was reconfigured to accommodate Tadao Ando's building with a circular floor plan. A reinforced concrete core was erected to house the toilet rooms, and an elliptical roof and aluminum-paneled curved walls were installed. The curved walls were surrounded by diagonally sloping louvers, and the space in-between was used as a passageway. Extra attention was paid to the detailing of the walls and louvers on site.

恵比寿公園トイレ **Ebisu Park Public Toilet** [P.32]

15枚の壁が複雑に組み合わされた片山正通のプランを、コンクリートの現場打ち込みでひと続きの躯体として立ち上げた。打設後に修正がきかないコンクリート打ち放し、かつ、内外とも全面的に杉板仕上げで、現場が注力した木目の表現の美しさも見所のひとつ。屋根面の施工記録写真は、それぞれの壁がぶつかり合う角度の複雑さがよくわかる。

Masamichi Katayama's plan, which consists of fifteen intricately composed walls, was erected as a single continuous structure using cast-in-place concrete. The exposed concrete, which cannot be modified after casting, used cedar board formwork throughout the interior and exterior, and the beauty of the grain patterns achieved through the careful attention of the site personnel is another highlight of the project. The construction documentation photos of the roof surface show the complexity of the angles at which the walls collide with each other.

裏参道公衆トイレ Urasando Public Toilet [P.142]

マーク・ニューソンのデザインによる、四隅が曲面のコンクリート打ち放しを再現するため、型枠も曲面パーツを用意し、割付を綿密に検討。なめらかな仕上がりをもたらす型枠の素材選びにもこだわった。全体に丸みを持たせた銅葺きの屋根は、下地となる木や鉄骨の骨組みの検証に3Dを駆使し、銅板は職方が現場で加工、手作業で打ち付けた。

Curved parts were prepared for the formwork, and the layout was carefully studied to realize Marc Newson's exposed concrete building with four rounded corners. The formwork materials were selected to provide a smooth finish. The wooden substrate and steel framework of the entirely rounded roof were analyzed using 3D technology, and specialist carpenters fabricated and applied the copper plates manually onsite.

恵比寿東公園トイレ Ebisu East Park Public Toilet [P.52]

施設ボリュームを分散して配置する計画で、土間コンクリート打設後の記録写真からは、その平面計画が見てとれる。槇文彦のデザインの特徴のひとつ、鉄板の曲面屋根は、宮城県気仙沼で造船技術を用いて製作。パーツごとに搬入され、現場で溶接した。建物全体の構造計算は、任意形状構造解析プログラムを用い、立体弾性解析を行っている。

The floor plan calls for a dispersed arrangement of facility volumes, evident from the photographic documentation taken after pouring the concrete for the earthen floor. One of the distinctive features of Fumihiko Maki's design, the curved roof made of steel plates, was fabricated using shipbuilding techniques in Kesennuma, Miyagi Prefecture. It was brought in individual parts and welded on site. The structural calculations of the entire building used an arbitrary shape structural analysis program to perform a three-dimensional elastic analysis.

写真提供：大和ハウス工業 Photo provided by Daiwa House Industry

Experiencing the Boundless Possibilities of Architecture

Keitaro Takebayashi
President of DAIWA LIFENEXT

Most buildings designed and constructed by Daiwa House Industry are completed entirely by our in-house engineers. Their job is to meet the client's requests and deliver a building while adhering to the cost and construction schedule. Since the company started from steel pipe structure construction and is committed to the philosophy of "industrialization of architecture," we usually do not seek a completely unexpected goal to land on.

The design and construction of "THE TOKYO TOILET" facilities had an entirely different feel from our long-cultivated style. From the outset, it was easy to imagine that the public toilets conceived by the sixteen creators, from Pritzker Prize-winning architects to creative directors active in the fashion and advertising industries, would be unique, and their construction would not be a simple matter.

Moreover, when the project team approached us, construction companies were fully engaged with the upcoming Tokyo Olympics and Paralympics, and the economy was improving thanks to economic stimulus measures such as monetary easing. They told us they had already approached other firms, but all of them had turned them down. Our company was also experiencing the same difficulties recruiting workers at the time. In fact, we were thinking of turning them down as well.

Nevertheless, we decided to accept the offer because the project deals with toilets, a place that everyone needs, and because it is rooted in the public spirit of "creating something everyone can use." At the heart of this project are the words of Nobuo Ishibashi, founder of Daiwa House Industry, who said, "We do what we do for the benefit of society, not to make money." We took the plunge and embarked on the project because we felt this project was in keeping with his spirit.

We were confident that the experience of participating in a project to create public facilities appreciated by the users and communicating with leading creators would be invaluable to the project members and our entire company. This project would be a once-in-a-lifetime opportunity, especially for young employees who have never worked with people outside the company.

We usually operate under a division system that is divided according to the building usage. However, we transcended the divisional boundaries for THE TOKYO TOILET project and assembled a project team across three departments: General Construction Business Division, Commercial Construction Business Division, and Apartment Business Division.

We selected participants through candidacy. Many employees ran for candidacy, mostly young people in their 20s and 30s, with some seasoned employees in senior positions. One of the notable characteristics of this project team was that more than half of the members were women, and many of the site supervisors were women.

In retrospect, our biggest challenge was probably the managing part of the construction management. Site conditions for public toilets are often more complex than those of general construction sites. Construction hours are limited in busy areas such as in front of train stations, and vehicles are not allowed in parks, making it difficult to bring in materials. Where playground equipment and sandboxes are close by, we needed to consider the possibility that children may enter the enclosure outside of construction hours. We had to prevent accidents by all means. Each site had specific concerns, and we consulted with Shibuya Ward, the Tokyo Metropolitan Government, and the police in advance and carefully planned the work to ensure safe operations.

Believing in the power of design to persevere in difficult situations

I feel that Fumihiko Maki's Ebisu East Park Toilet was the first hurdle in the design and construction of THE TOKYO

Keitaro Takebayashi

Born in 1969. After working as General Manager of the General Construction Division and Head of the Tokyo Head Office of Daiwa House Industry Co., Ltd., he was appointed Executive Officer of the company in 2019. Since April 2022, he has been President of DAIWA LIFENEXT Co., Ltd.

TOILET facilities, which we struggled to overcome and vowed to continue to the end. The steel roof combining many curves was fabricated using shipbuilding techniques at a factory in Kesennuma, Miyagi Prefecture, and were welded on site. We had never worked with shipbuilders before and were first worried about the cost and process. But we had to do it anyway, so we took the plunge.

Nao Kubota's Higashi Sanchome Public Toilet is surprisingly compact, and it made us wonder, "Is this really the right place?" upon seeing the site for the first time. However, once completed, the three toilets, including the universal toilet, fit neatly into the site and offer highly comfortable spaces. When we saw the power of this design, we realized that we could never say impossible, whatever the budget, time, or physical constraints we might face. Over the next five years, we worked to realize "public toilets like never before," returning again and again to the project's purpose of embracing the creators' plans and making them a reality.

Many challenges will turn into invaluable assets

After successfully completing the seventeen facilities of THE TOKYO TOILET, we asked the project team members about the highlights of the toilet facilities they were in charge of once again, and they each spoke passionately about this project. In large buildings such as logistics facilities, each person is usually assigned to a specific process or part of the building. On the other hand, public toilets are small in scale, allowing one to see the entire facility from start to finish, and eventually develop a strong commitment to the project. More importantly, it was a project that they voluntarily participated in. The members were highly motivated from start to finish, and this high level of motivation was perhaps our greatest strength in this project.

THE TOKYO TOILET" has several buildings made of reinforced concrete. We painstakingly finished the exposed concrete exterior as beautifully as possible by increasing reinforcing bars, modifying the material quality, and providing new formwork. Some of them are cast-in-place concrete with rounded corners, and the ingenious allocation of the formwork helped realize the three-dimensional curved surfaces.

Each person in charge also took great care in coating and painting. The steel plates were coated with a thermal insulating paint, and the white exterior walls were coated with a photocatalytic paint that uses sunlight to decompose dirt for easy cleaning.

The copper roof of Marc Newson's "Urasando Public Toilet" is another memorable example of the construction process. The copper roof he envisioned was unique and different from the standards and concepts of Japanese temples and shrines and presented a new challenge to the carpenters specializing in this field. The size of the copper plates was irregular, and we used 3D technology to study the wood substrate and steel framework. We also carefully conducted on-site machining and scraping to recreate the rounded shapes.

In the later stages of construction, the staff in charge of the project seemed more confident as they continued to meet challenges until they were satisfied with the results. We are proud that our project members received praise for their tenacity and high level of perfection, and we firmly believe that we successfully embodied the creators' visions.

We often had to jump over a number of unimaginable and high hurdles during our involvement in the project. Yet the experience of negotiation, coordination, design, and construction that culminated in the realization of the project became an invaluable asset to Daiwa House Industry. We hope THE TOKYO TOILET facilities will be used with care and contribute to society. We invite you to visit the facilities and experience the endless possibilities of public toilets and architecture, which differ from site to site.

SPECIAL CONTRIBUTION

忙しい東京の中の静かな場所

ヴィム・ヴェンダース
映画監督

私の母語ドイツ語では、かつてトイレに愛称があった。ダス・シュティレ・エルトヒェン(das stille Örtchen)。ひとまず「例の静かな場所」と訳せるだろう。とはいえこの言い方には、翻訳では伝わらないものがこもっている。トイレという、人間には欠くべからざる場所を、情愛をこめて語る言葉だからだ。「ここへ来ればあなたは安らかな気持ちになれる。しっかり護られて、心地よく過ごせて、落ち着いて考えをまとめる余裕もできて、真に一人になれる。ここは世のあわただしさからの避難所なんだよ」。Ort(場所)を慈しんでchenを付けるだけで、これだけの思いがこもるのだ。同じくPlatz(場)を慈しんでPlätzchenとすることもできる。英語では「クッキー」の意だが、本来は「くつろげる場」ということであり、これに同じくstill(静かな)を加えてein stilles Plätzchenとすれば「安全な避難所」である。

今日、トイレの古い呼び名を使う人はもういない。私の場合、祖母がいつも使っていたのでいまだ頭にあるわけだが、日常的に用いる言葉とはいかない。何と言っても、いまの世の中、トイレにはもうそのような含意がないのだ。しかし、「THE TOKYO TOILET」プロジェクトと出会い、その奇跡のような衛生の神殿を初めて目にしたとき、まっさきに頭に浮かんだのはこのずっと昔の言葉だった。どのトイレも、あの古き良き言い回しがぴったりなのだ(もっとも私の祖母は、これほど快適な「静かな場所」に座るなんて考えもしなかっただろうが……)。

これらTTTのトイレは、我々がふだん知るトイレとはまるで違う。きわめてモダンな、先端技術の粋とも言うべき衛生の場である。とはいえ、そこから発散されているのは、まさに私の祖母が言わんとしていたことだ。それらはいわば、祖母が用足しに行ったときに抱いた欲求を、現代の言語に翻訳している。むろん言語といっても建築の言語だが、決して企業的であったり、プレハブであったりはしない。17のささやかな名作はそれぞれが唯一無二であり、二つとして同じものはない。現代の建築家たちが、それなりの自由と資力を得られればどれだけのものを夢見てみせるか、これらは証している。「あなたが夢見る理想の公共トイレをデザインせよ!」とは、およそよくある使命とは言えない。むろん建築家はトイレを設計するし、時には美しいトイレだって設計する。だがそれはつねに建物の一部という文脈に属している。独立したトイレを作ることはまずない。

それゆえ、これら「東京の中の静かな場所」はきわめて特別なものになっている。それぞれが都市の風景の中の、ひとつしかない空間を占めている。大半は公園や木の下に作られていて、そうした位置だけを見ても、よく配慮された、貴重な場であることがわかる。それぞれがそこに、まさに建っている場に「属している」ように思える。どれもがその場にあつらえて作られた、その場にあるからこそ成り立つ芸術作品だからだ。実際に見てみるとそのことが実感できるし、中に入ればもっとよくわかる。これは丁寧に、きちんと考えて作られた場所なのだという、きわめてはっきりした感覚。それは力強いメッセージであり、使う者にも伝染する。これら清潔で現代的な便器や洗面台を使うとき、あなたは自分が人間として重んじられている気になれる。かけがえのない存在として扱ってもらっている思いがする。私の祖母の言葉が伝えていたとおり、自分が慈しまれている気持ちになれるのだ。

(訳 柴田元幸)

訳者より　ドイツ語に関して片山耕二郎さんにご教示いただいた。深く感謝する。

ヴィム・ヴェンダース
1945年ドイツ・デュッセルドルフ生まれ。脚本家、監督、プロデューサー、写真家、作家としての顔をもち、長編映画のほか革新的なドキュメンタリー映画の作り手としても知られる。『パリ、テキサス』(84年)、『ベルリン・天使の詩』(87年)、『ブエナ・ビスタ・ソシアル・クラブ』(99年)、『セバスチャン・サルガド／地球へのラブレター』(2014年)など作品および受賞歴多数。「THE TOKYO TOILET」プロジェクトの一環として誕生した映画『PERFECT DAYS』(2023年)の監督、共同脚本。

Quiet Spots in Busy Tokyo

Wim Wenders
Film director

In my language, in German, toilets had a nickname. A toilet was called "das stille Örtchen" which translates vaguely into "the quiet spot". Then again, the expression escapes any translation, as it is really a very endearing way to talk about that place of human necessity. It really says: "Here you can be in peace. You'll be well protected and comfortable, you have time to gather your thoughts and be really by yourself. This is a refuge from the buzz of the world." All this resonates in the diminutive form of the word "Ort". If you want, "Plätzchen" is really a "cosy place", and combining it with "still", the expression really defines a "safe haven".

Nowadays, nobody uses that old saying any more. My grandmother always used it, that's how I still know it. But it's not part of my everyday vocabulary. Simply because in my time toilets don't have the above connotation any more. However, that ancient term was the first thought that came to my mind when I encountered THE TOKYO TOILET project and saw these miraculous sanitary temples for the first time. They all fitted that lovely old saying, even if my grandmother would have never dreamed of ever sitting on her "quit spot" in such comfort.

These TTT toilets are in no way embodiments of toilets as we know them. They are extremely modern "state of the art" sanitary places, yet they radiate exactly what my grandmother was referring to. They translate, so to speak, her desires when she went to the loo into a modern language. This is an architectural language, sure, but in no way a corporate or prefabricated one. All of these 17 little masterpieces are unique, none is like the other. They show what contemporary architects are able to dream up, if you give them the liberty and the means to do so. "Design your ideal vision of a public toilet!" is certainly not a common mission. Architects plan toilets, sometimes beautiful ones, but always in the context of a building. They don't build toilets on their own.

But that is exactly what makes these "quiet spots in Tokyo" so special. Each of them fits a unique space in the urban landscape. Most of them found their places in parks and under trees. Their locations alone mark them as thoughtful and precious. Each of them seems to "belong there", exactly where they stand, as they are all made to measure, site-specific pieces of art. Which is something you feel when you see them and even more so, when you enter. That very distinct feeling that they are made with care and consideration is a powerful message that rubs off on the user. And when you use these clean and modern toilets, wash basins and other utilities, you feel respected yourself, as a human being. You feel you are treated as unique. You feel valued, like in my grandmother's expression.

Wim Wenders
Born in 1945 in Düsseldorf, Germany. Wim Wenders is a screenwriter, film director, producer, photographer, and writer widely known for his feature films and innovative documentary films. His numerous films and awards include "Paris Texas" (1984), "Der Himmel über Berlin" (1987), "Buena Vista Social Club" (1999), and "The Salt of the Earth" (2014). He directed and co-wrote the film "PERFECT DAYS" (2023), co-produced as part of THE TOKYO TOILET project.

DRAWINGS AND SPECIFICATIONS

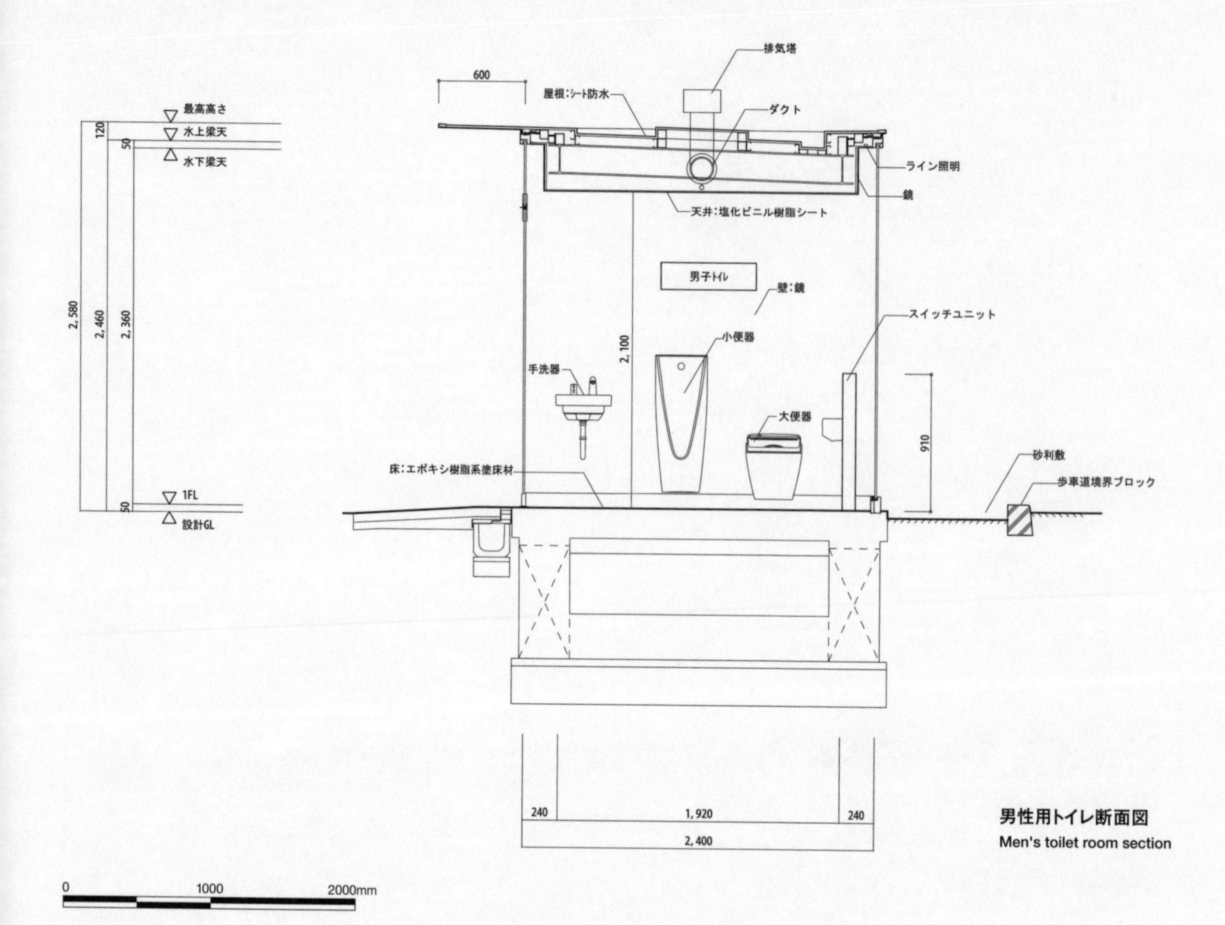

男性用トイレ断面図
Men's toilet room section

代々木深町小公園トイレ P.18
YOYOGI FUKAMACHI MINI PARK PUBLIC TOILET

はるのおがわコミュニティパークトイレ P.22
HARU-NO-OGAWA COMMUNITY PARK PUBLIC TOILET

【所在地】［代々木深町小公園］東京都渋谷区富ヶ谷1-54-1／［はるのおがわコミュニティパーク］東京都渋谷区代々木5-68-1
【デザイン】坂茂建築設計／坂 茂 菅井啓太 梶原慧太
【設計・施工】大和ハウス工業

【規模】［代々木深町小公園］
敷地面積 134.33m²／建築面積 14.16m²
延床面積 14.16m²／建蔽率 10.55%
(許容: 60%)［はるのおがわコミュニティパーク］
敷地面積 152.39m²／建築面積 14.16m²
延床面積 14.16m²／
建蔽率 9.30%(許容: 60%)

【寸法】最高高 2,580mm／
軒高 2,460mm／天井高 2,100mm
【敷地条件】第二種中高層住居専用地域
準防火地域 第二種高度地区
【構造】鉄骨造
【施工期間】2020年3月〜7月

【外部仕上げ】
屋根: シート防水
外壁: ガラス: 強化ガラス+カラーフィルム+瞬間調光フィルム／サッシ: ステンレス鏡面
外構: 砂利 インターロッキングブロック
【内部仕上げ】
床: エポキシ樹脂系塗床 壁: 鏡
天井: 塩化ビニル樹脂シート

【主な使用機器】(TOTO)
壁掛大便器: UAXC3CS1／ウォシュレット(エコリモコン): TCF5840AUPR／
壁掛小便器: XPU21A／
壁掛洗面器: LS721CM／
台付自動水栓: TENA41A／
コンパクトオストメイトパック: UAS81RSB2NW
ベビーシート: YKA25R／
ベビーチェア: YKA15R

Location
[Yoyogi Fukamachi Mini Park Public Toilet] 1-54-1 Tomigaya, Shibuya-ku, Tokyo
[Haru-no-Ogawa Community Park Public Toilet] 5-68-1 Yoyogi, Shibuya-ku, Tokyo
Design
Shigeru Ban Architects/ Shigeru Ban, Keita Sugai, Keita Kajihara
Design and construction
Daiwa House Industry

Size
[Yoyogi Fukamachi Mini Park]
Site area 134.33 m²/
Building area 14.16 m²/
Total floor area: 14.16 m²/
Building coverage ratio 10.55%
(Maximum allowable ratio 60%)
[Haru-no-Ogawa Community Park]
Site area: 152.39 m²/
Building area 14.16 m²/
Total floor area: 14.16 m²/
Building coverage ratio 9.30%
(Maximum allowable ratio:60%)
Dimension
Maximum building height 2,580 mm/
Eave height 2,460 mm/
Ceiling height 2,100 mm
Site condition
Category 2 Medium-to-high-rise Exclusive Residential District,
Quasi-fire Prevention District,
Category 2 Height Control District

Structure
Steel construction
Construction period
March-July 2020

Exterior finishes
Roof: Membrane waterproofing
Exterior wall/ glazing: Tempered glass with tinted film and instant-dimming film with stainless steel window frames
Exterior: Pebble and interlocking block
Interior finishes
Floor: Epoxy resin floor coating
Wall: Mirror / Ceiling: PVC sheet

Sanitary fixtures used
(TOTO products)
Toilet: UAXC3CS1/
Washlet (with Eco Remote Controller): TCF5840AUPR/
Urinal: XPU21A/
Wash basin: LS721CM/
Automatic faucet: TENA41A/
Ostomate sink: UAS81RSB2NW/
Baby changing station: YKA25R/
Baby seat: YKA15R

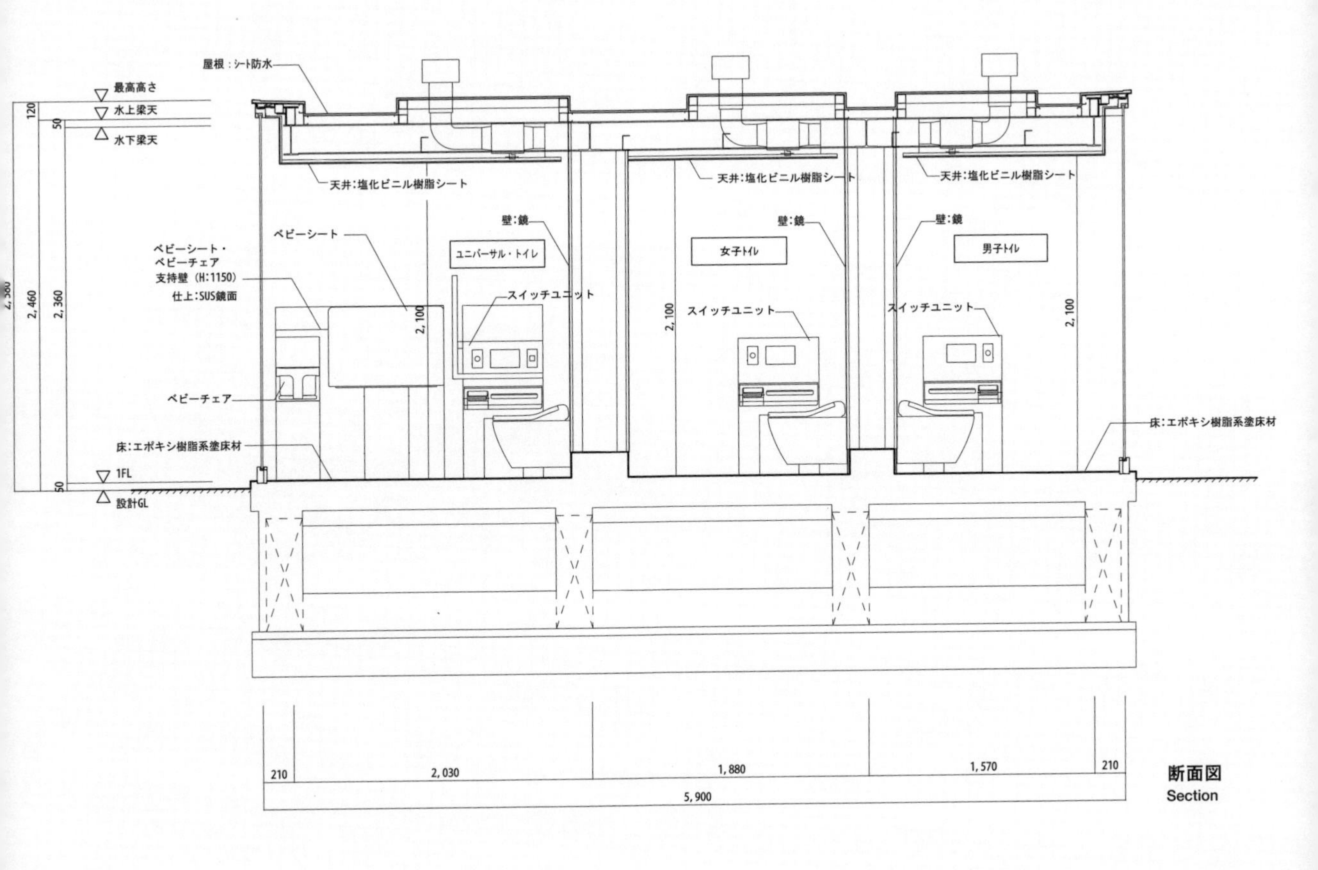

断面図
Section

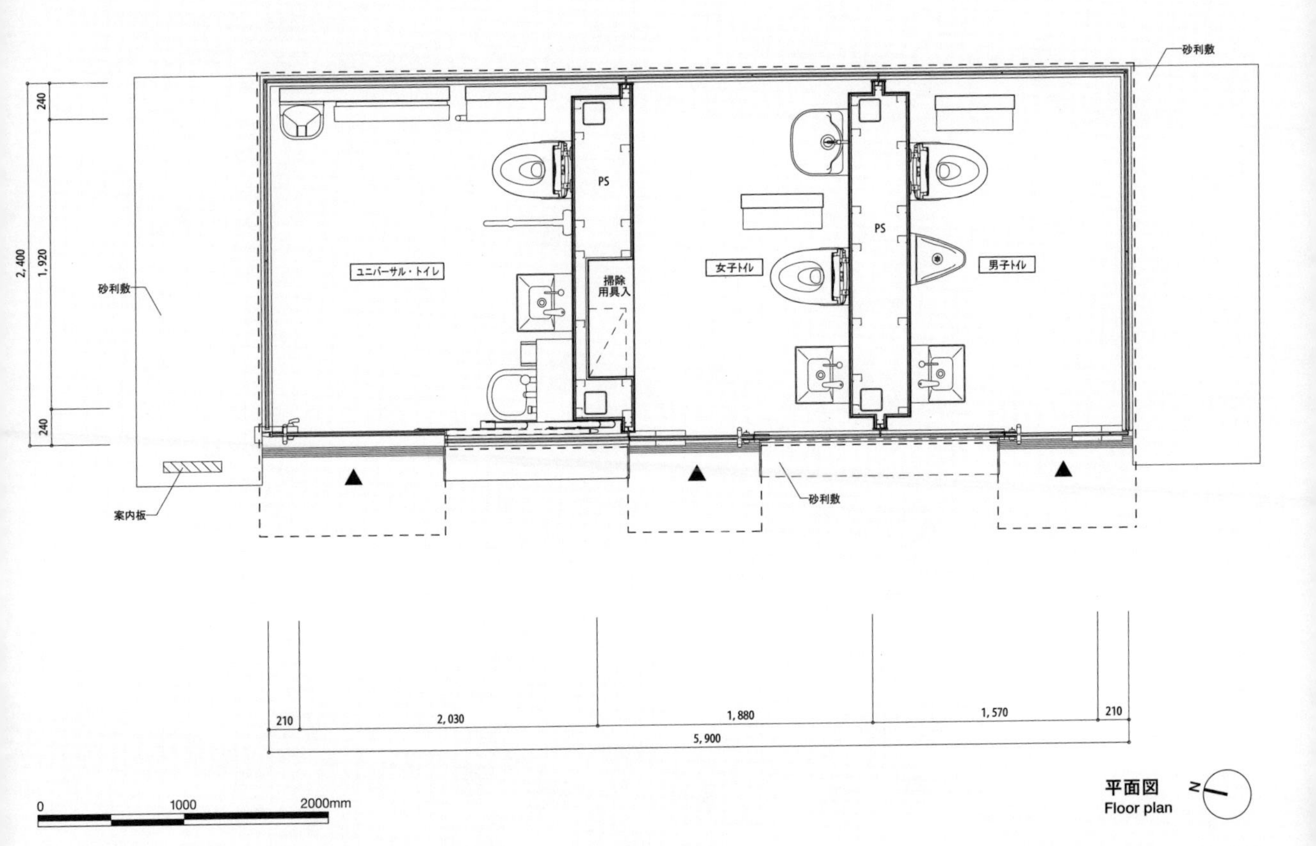

平面図
Floor plan

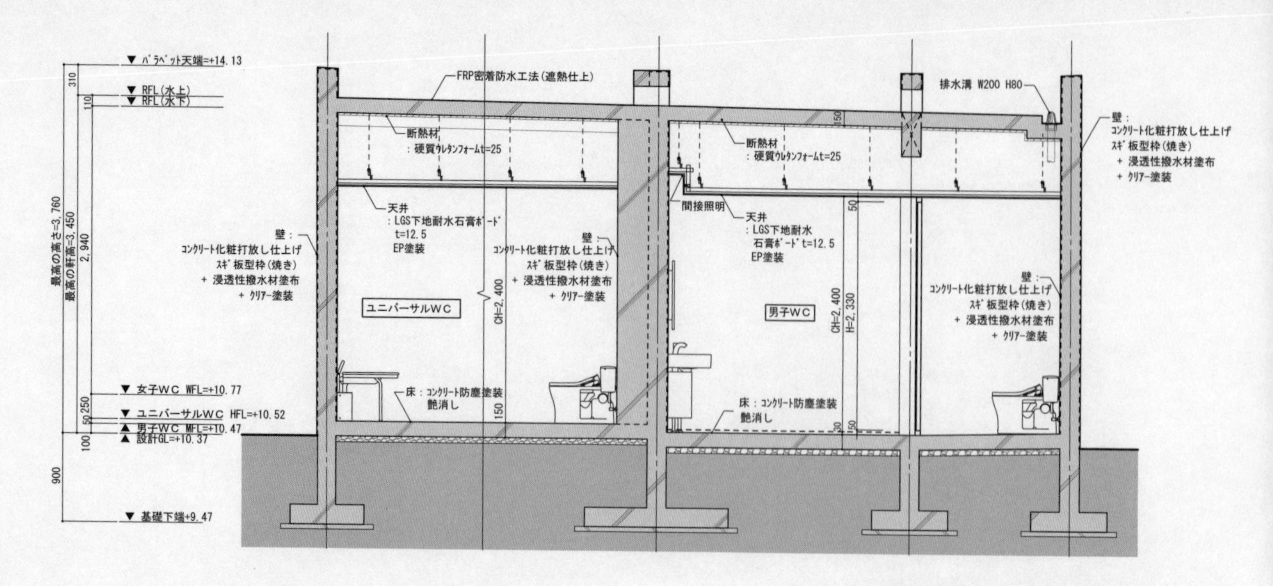

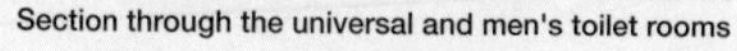

ユニバーサル・トイレ 男性用トイレ断面図
Section through the universal and men's toilet rooms

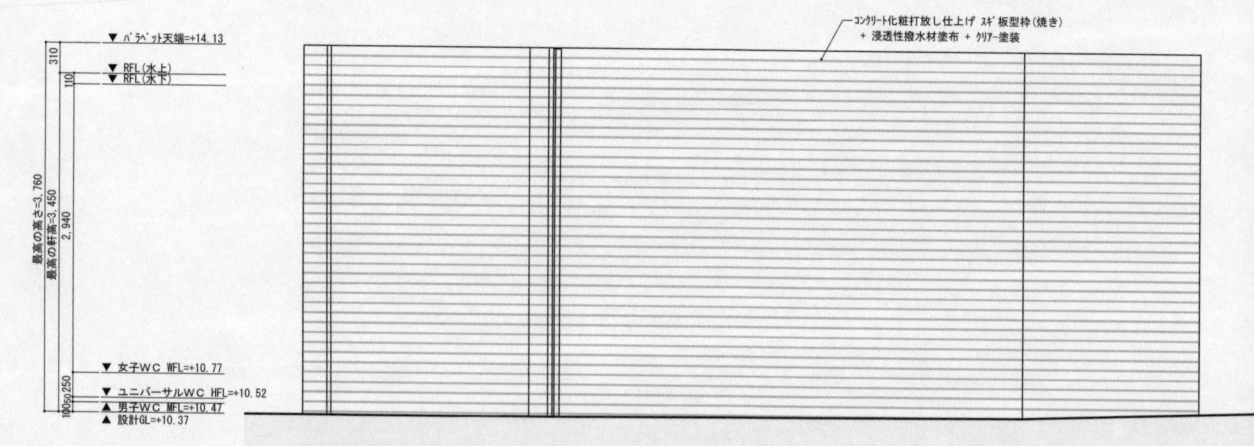

立面図
Elevation

0 1000 2000mm

恵比寿公園トイレ P.32

EBISU PARK PUBLIC TOILET

【所在地】東京都渋谷区恵比寿西1-19-1
【デザイン】ワンダーウォール®/片山正通
タライエ ファラザネ* 中村浩樹
上原紗保里* 栗原秀圭(*元所員)
【設計・施工】大和ハウス工業

【規模】敷地面積 198.41m²/
建築面積 37.52m²/延床面積 37.52m²
建蔽率 18.91%(許容: 建築基準法80%
都市公園法公園全体面積の2%)
【寸法】最高高 3,760mm/
軒高 3,500mm
【敷地条件】
第二種中高層住居専用地域 準防火地域
第三種高度地区・商業地域 防火地域
【構造】鉄筋コンクリート造
【施工期間】2020年2月~7月

【外部仕上げ】
屋根: FRP密着防水工法
外壁: コンクリート化粧打ち放し仕上げ
スギ板型枠(焼き)撥水剤塗布
開口部: スチール水性ローバル仕上げ
外構: コンクリート防塵塗装
【内部仕上げ】
床: コンクリート防塵塗装
壁: コンクリート化粧打ち放し仕上げ
スギ板型枠(焼き)撥水剤塗布
天井: PB t=12.5mm+EP塗装

【主な使用機器】(TOTO)
パブリックコンパクト便器: CFS497BC/
ウォシュレット(エコリモコン): TCF5840AUPR
自動洗浄小便器: UFS910JS/
壁掛洗面器: L710CM/
台付自動水栓: TENA12F/オストメイト
対応マルチパック: XPSA73C71WW/
収納式多目的シート: EWC520ARR/
ベビーチェア: YKA15R、YKA16R/
ベビーシート: YKA25R

Location
1-19-1 Ebisunishi, Shibuya-ku, Tokyo
Design
Wonderwall®/ Masamichi Katayama, Farzaneh Talaie*, Hiroki Nakamura, Saori Uehara*, Hidekado Kurihara
(*Former staff members)
Design and construction
Daiwa House Industry

Size
Site area 198.41 m²/
Building area 37.52 m²/
Total floor area 37.52m²/ Building coverage ratio 18.91%
(Maximum allowable ratio: Building Standards Act - 80%/
Urban Park Act - 2% of total park area)
Dimension
Maximum building height 3,760 mm/
Eave height 3,500 mm
Site condition
Category 2 Medium-to-high-rise Exclusive Residential District, Quasi-fire Prevention District, Category 3 Height Control District, Commercial District, Fire Prevention District
Structure
Reinforced concrete construction
Construction period
February - July 2020

Exterior finishes
Roof: FRP adhesive waterproofing
Exterior wall: Exposed fair-faced concrete finish using water repellent coated charred cedar formwork
Opening: Water-based Roval paint
Exterior: Dustproof coated concrete
Interior finishes
Floor: Dustproof coated concrete
Wall: Exposed fair-faced concrete finish using water repellent coated charred cedar formwork
Ceiling: Plasterboard t=12.5 mm, emulsion paint

平面図
Floor plan

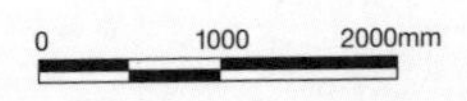

Sanitary fixtures used
(TOTO products)
Toilet: CFS497BC/
Washlet (with Eco Remote Controller):
TCF5840AUPR/
Urinal: UFS910JS/ Wash basin: L710CM/
Automatic faucet: TENA12F/
Ostomate sink: XPSA73C71WW/
Fold-down nursing care bed:
EWC520ARR/
Baby seat: YKA15R, YKA16R/
Baby changing station: YKA25R

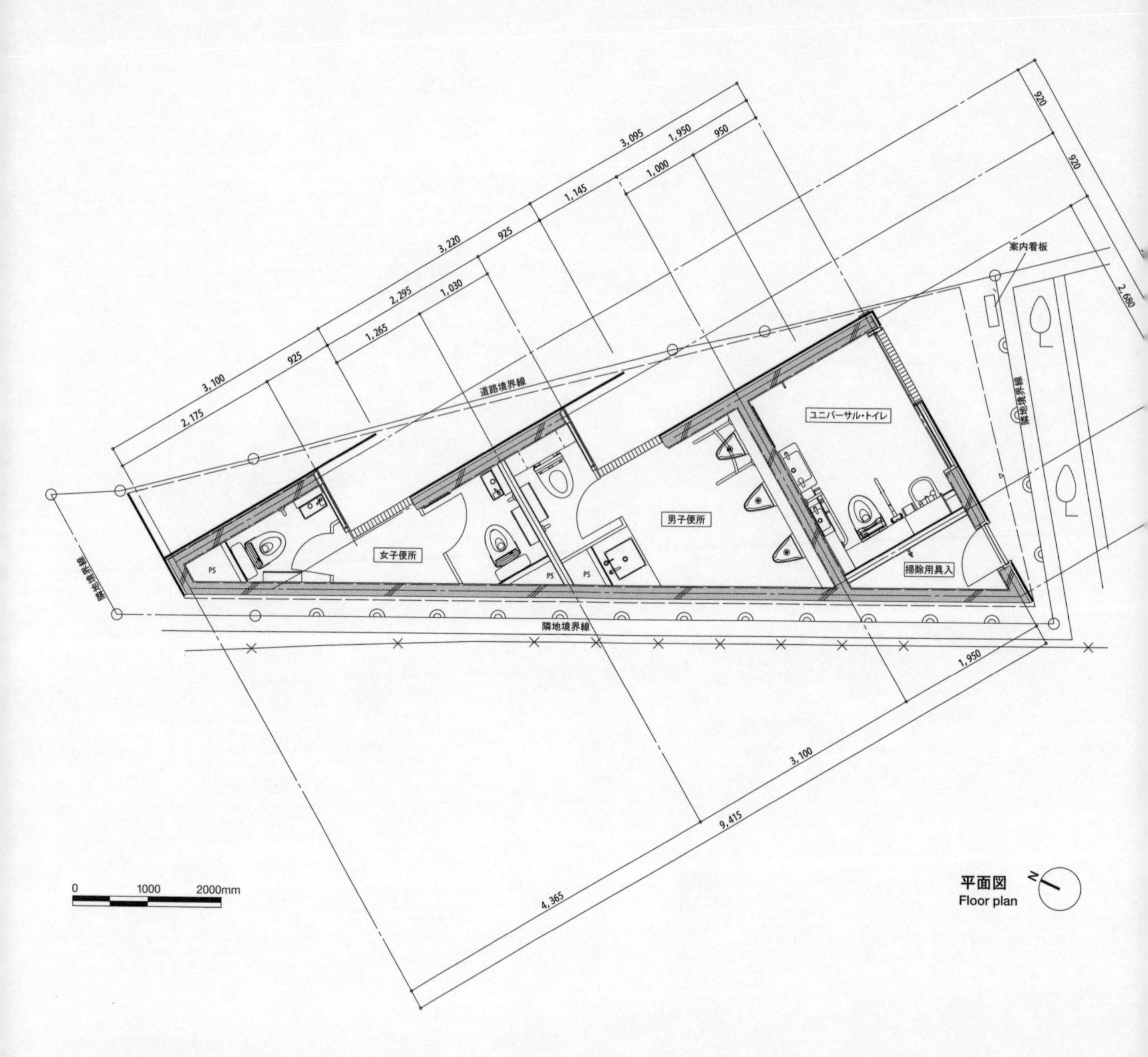

東三丁目公衆トイレ P.42

HIGASHI SANCHOME PUBLIC TOILET

【所在地】 東京都渋谷区東3-27-1
【デザイン】 田村奈穂
【設計・施工】 大和ハウス工業

【規模】 敷地面積 36.85m²／
建築面積 19.25m²／
延床面積 18.84m²／
建蔽率 52.24%（許容: 80%）
【寸法】 最高高 3,622mm／
軒高 3,350mm／天井高 2,400mm
【敷地条件】 商業地域 防火地域
60m高度地区
【構造】 壁式鉄筋コンクリート造
【施工期間】 2020年5月〜8月

【外部仕上げ】
屋根・外壁: 構造用鋼板＋
超耐久型フッ素樹脂系遮熱塗料
外構: アスファルト舗装
【内部仕上げ】
床: エポキシ樹脂系塗床
壁: コンクリートの上、EP-G塗装
天井: 珪酸カルシウム板 t=8mmの上、
EP-G塗装

【主な使用機器】（TOTO）
壁掛大便器: UAXC3CS1／
パブリックコンパクト便器: CFS497BC／
ウォシュレット（エコリモコン）: TCF5840AUPR
壁掛小便器: XPU21A／
洗面器: LS715／手洗器: L50D／
台付自動水栓: TENA12AW、TENA12B／
コンパクト多機能トイレパック:
UADAK21R1A1ASN2WA

Location
3-27-1 Higashi, Shibuya-ku, Tokyo
Design
Nao Tamura
Design and construction
Daiwa House Industry

Size
Site area 36.85 m²/
Building area 19.25 m²/
Total floor area 18.84m²/
Building coverage ratio 52.24%
(Maximum allowable ratio: 80%)
Dimension
Maximum building height 3,622 mm/
Eave height 3,350 mm/
Ceiling height 2,400 mm
Site condition
Commercial District,
Fire Prevention District,
60 m-Height Control District
Structure
Reinforced concrete wall construction
Construction period
May - August 2020

Exterior finishes
Roof / exterior wall: Steel plate for general construction purposes with super-durable fluoropolymer thermal insulation coating
Exterior: Asphalt paving
Interior finishes
Floor: Epoxy resin floor coating
Wall: Concrete with gloss emulsion paint finish
Ceiling: Calcium silicate board t=8 mm with gloss emulsion paint finish

Sanitary fixtures used
(TOTO products)
Toilet: UAXC3CS1/
Toilet: CFS497BC/
Washlet (with Eco Remote Controller): TCF5840AUPR/
Urinal: XPU21A/
Wash basin: LS715/
Hand wash basin: L50D/
Automatic faucet: TENA12AW, TENA12B/
Wheelchair accessible toilet unit: UADAK21R1A1ASN2WA

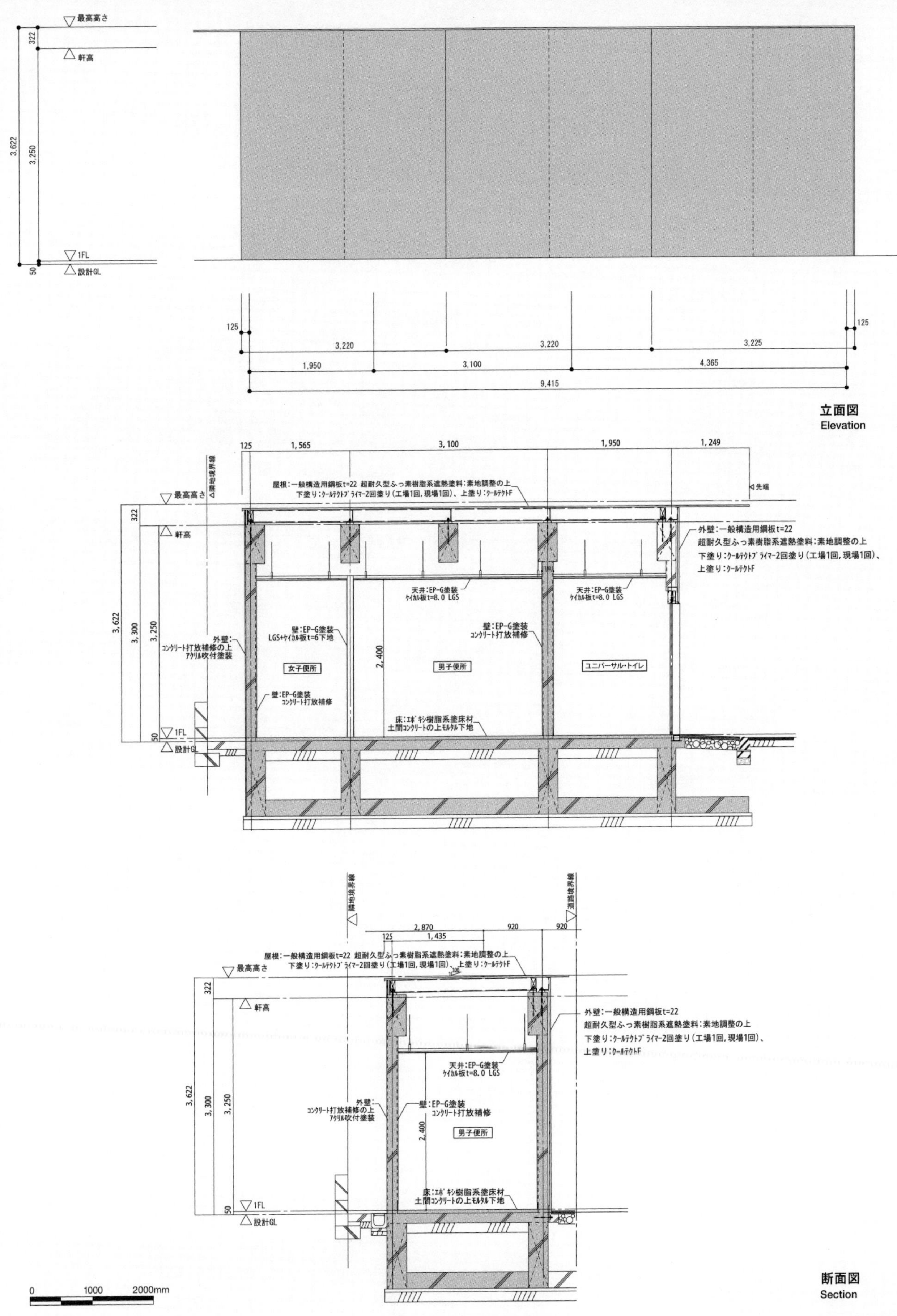

立面図
Elevation

断面図
Section

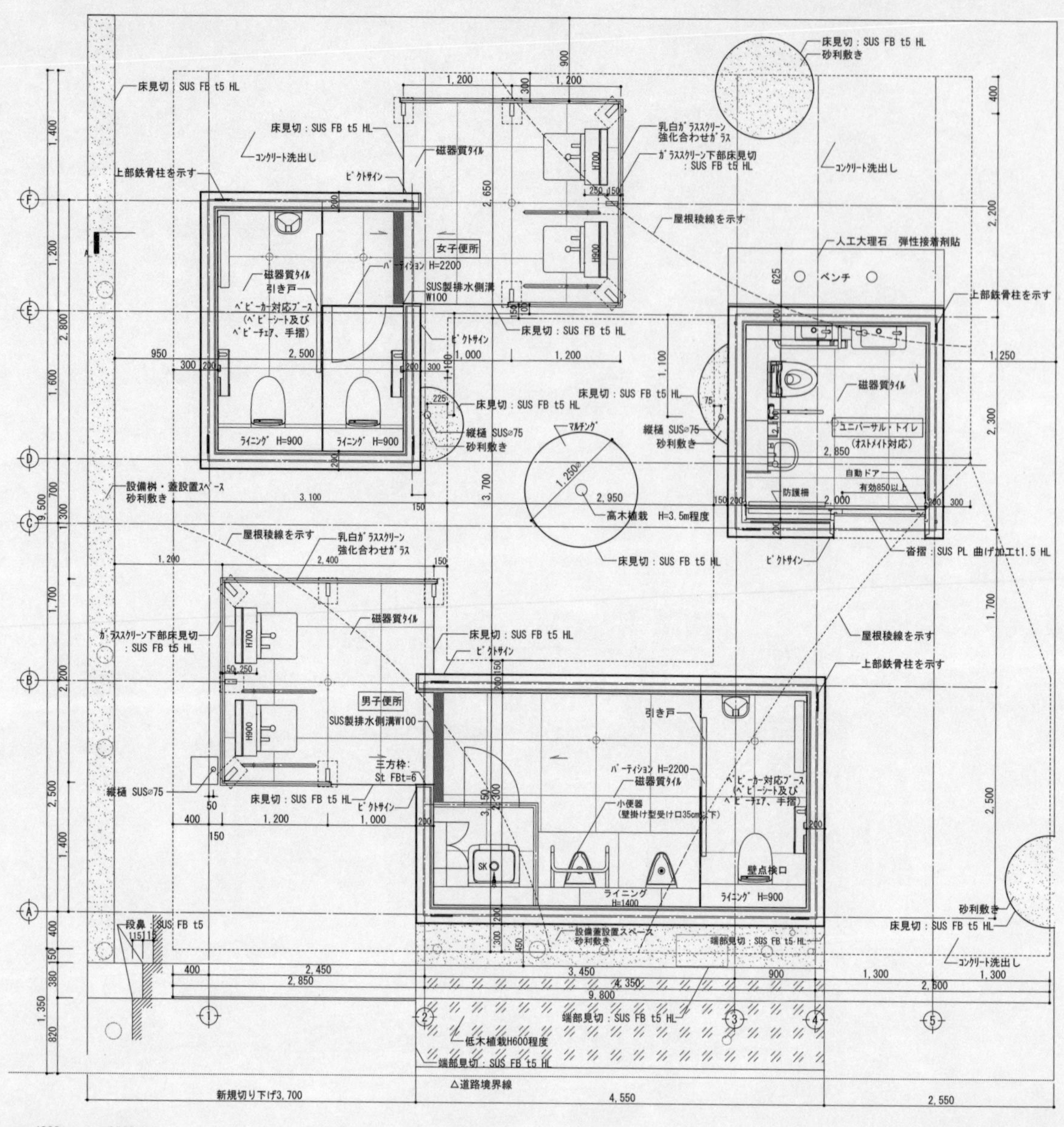

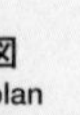

平面図
Floor plan

恵比寿東公園トイレ P.52
EBISU EAST PARK PUBLIC TOILET

【所在地】東京都渋谷区恵比寿1-2-16
【デザイン】槇総合計画事務所／槇文彦
長谷川龍友* 松田浩幸 西尾大河
吉野わか子*(*元所員)
【構造デザイン協力】KAP／桐野康則
【設計・施工】大和ハウス工業

【規模】敷地面積 215.18m²／
建築面積 48.80m²／
延床面積 42.85m²／
建蔽率 22.68%(許容: 80%)
【寸法】最高高 4,111mm／
軒高 2,720mm
【敷地条件】商業地域 防火地域
60m高度地区
【構造】鉄筋コンクリート造 一部鉄骨造

【施工期間】2020年4月～7月
【外部仕上げ】
屋根: 曲面鋼板 t=16mm厚膜型ポリシロキサン塗装
外壁: アクリル樹脂プラスター吹付
開口部: ガラススクリーン
外構: コンクリート洗い出し
ベンチ: 人工大理石
【内部仕上げ】
床: 磁器質タイル
壁: 磁器質モザイクタイル
天井: 厚膜型ポリシロキサン塗装

【主な使用機器】(TOTO)
壁掛大便器: UAXC3CS1／
ウォシュレット(エコリモコン): TCF5840AUPR
壁掛小便器: XPU21A／
壁掛ハイバック洗面器: LSA135FA／
壁付自動水栓: TENA125A／
コンパクト多機能トイレパック:
UADAK21R1A1ASN2WA／
ベビーチェア: YKA15R／
ベビーシート: YKA25R

Location
1-2-16 Ebisu, Shibuya-ku, Tokyo
Design
Maki And Associates/
Fumihiko Maki, Tatsutomo Hasegawa*, Hiroyuki Matsuda, Taiga Nishio, Wakako Yoshino* (*Former staff members)
Structural design cooperation
KAP/ Yasunori Kirino
Design and construction
Daiwa House Industry
Size
Site area 215.18 m²/
Building area 48.80 m²/
Total floor area 42.85 m²/
Building coverage ratio 22.68%
(Maximum allowable ratio: 80%)
Dimension
Maximum building height 4,111 mm/
Eave height 2,720 mm
Site condition
Commercial District,
Fire Prevention District,
60 m-Height Control District
Structure
Reinforced concrete construction, partially steel construction

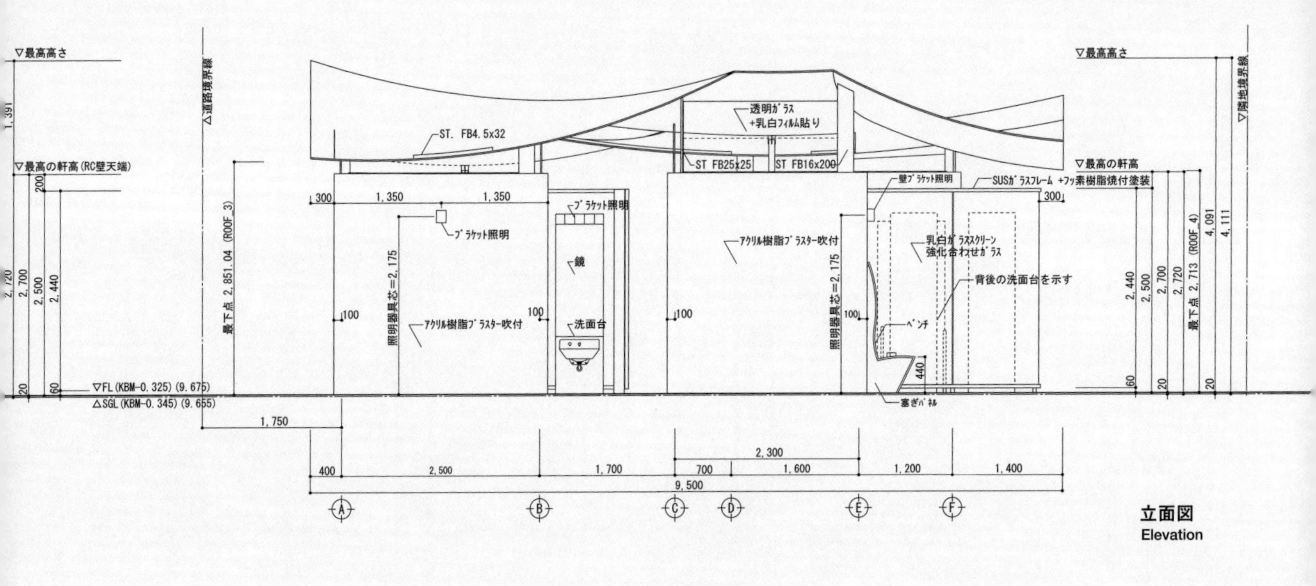

立面図
Elevation

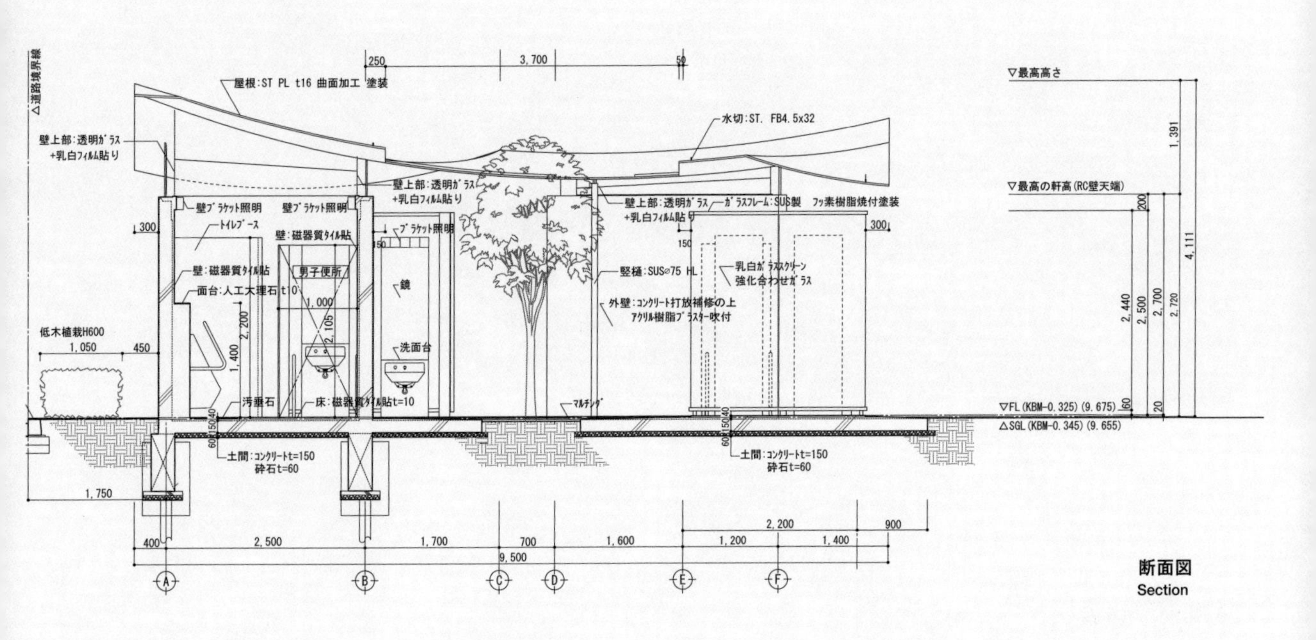

断面図
Section

Construction period
April - July 2020
Exterior finishes
Roof: Curved steel plate t=16 mm with thick film polysiloxane coating
Exterior wall: Acrylic resin plaster spraying inorganic coating
Opening: Glass screen
Exterior: Exposed-aggregate finish concrete
Bench: Synthetic marble
Interior finishes
Floor: Porcelain tile
Wall: Porcelain mosaic tile
Ceiling: Thick film polysiloxane coating

Sanitary fixtures used
(TOTO products)
Toilet: UAXC3CS1/
Washlet (with Eco Remote Controller): TCF5840AUPR/
Urinal: XPU21A/ Wash basin: LSA135FA/
Automatic faucet: TENA125A/
Wheelchair accessible toilet unit: UADAK21R1A1ASN2WA/
Baby seat: YKA15R/
Baby changing station: YKA25R

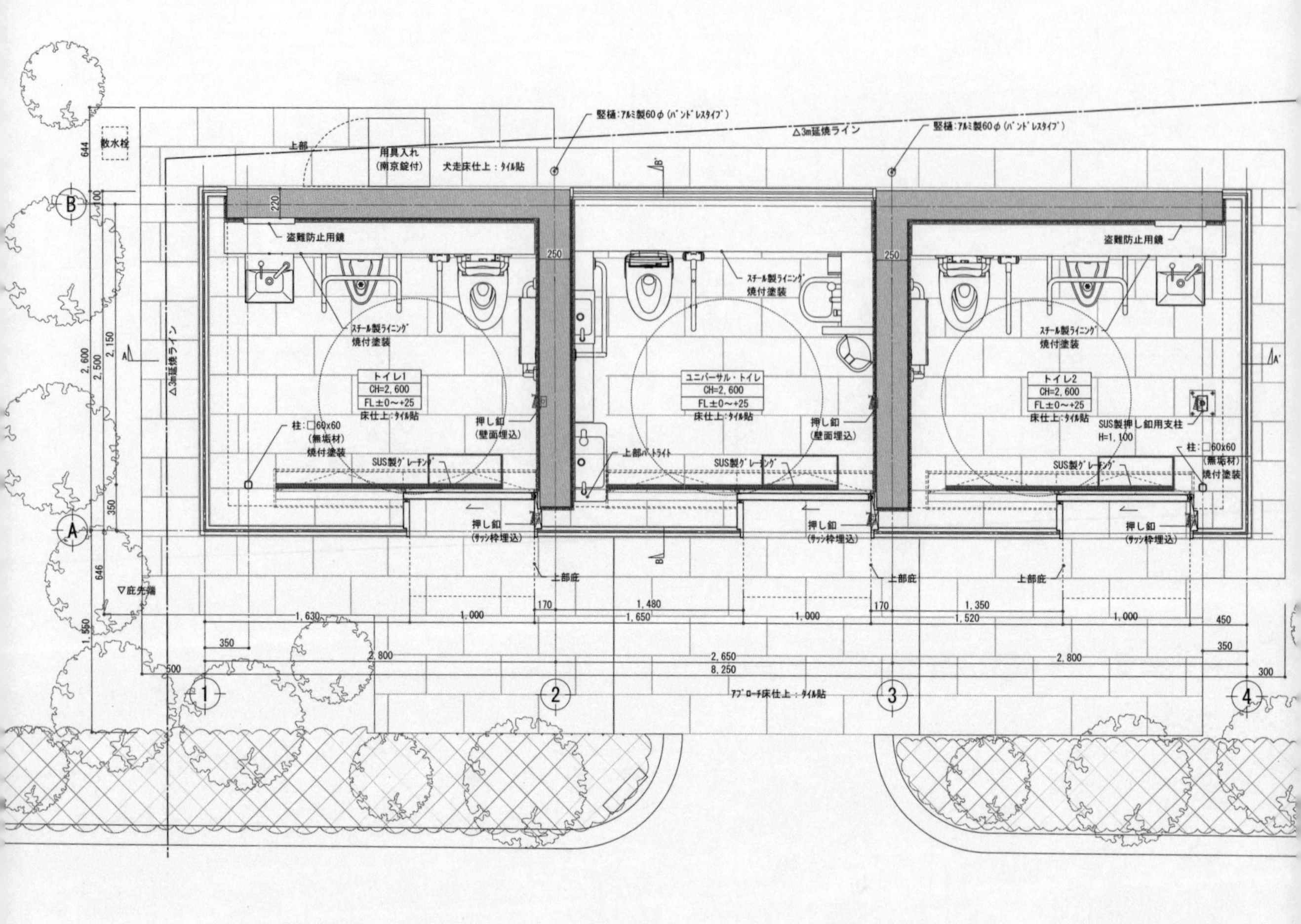

0 1000 2000mm

平面図
Floor plan

西原一丁目公園トイレ P.62

NISHIHARA ITCHOME PARK PUBLIC TOILET

【所在地】東京都渋谷区西原1-29-1
【デザイン】坂倉建築研究所／坂倉竹之助
大木健逸　大森祥司　小林大祐(元所員)
【設計・施工】大和ハウス工業

【規模】敷地面積　226.30m²／
建築面積　21.45m²／
延床面積　21.45m²／
建蔽率　9.47%(許容: 70.99%)
【寸法】最高高　3,240mm／
軒高　3,050mm／階高　3,050mm／
天井高　2,600mm(ユニバーサル・トイレ)
【敷地条件】商業地域　第一種住居地域
防火地域　50m高度地区
都市高速鉄道内
【構造】鉄筋コンクリート造　一部鉄骨造
【施工期間】2020年3月～8月

【外部仕上げ】
屋根: アルミハニカムパネル
外壁: ベルアートF
開口部: ステンレスサッシ　アルミサッシ
外構: インターロッキングブロック
【内部仕上げ】
床・壁: 磁器質タイル
天井: NAD塗装

【主な使用機器】(TOTO)
壁掛大便器: UAXC3CS1／ウォシュレット
(エコリモコン): TCF5840AUPR／
壁掛小便器: XPU21A／
壁掛洗面器: LS721／
台付自動水栓: TENA12AW／
コンパクトオストメイトパック:
UAS81RSB2NW／
ベビーチェア: YKA16R

Location
1-29-1 Nishihara, Shibuya-ku, Tokyo
Design
Sakakura Associates/
Takenosuke Sakakura,
Kenichi Ooki, Shoji Omori,
Daisuke Kobayashi (Former staff member)
Design and construction
Daiwa House Industry

Size
Site area 226.30 m²/
Building area 21.45 m²/
Total floor area 21.45m²/ Building coverage ratio 9.47%
(Maximum allowable ratio: 70.99%)
Dimension
Maximum building height 3,240 mm/
Eave height 3,050 mm/
Floor height 3,050 mm/
Ceiling height 2,600 mm (Universal toilet)
Site condition
Commercial District, Category 1 Residential District, Fire Prevention District, 50 m-Height Control District, site within the urban rapid transit system
Structure
Reinforced concrete construction, partially steel construction

Construction period
March - August 2020

Exterior finishes
Roof: Aluminum honeycomb panel
Exterior wall: Bell Art F
Opening: Stainless steel window frames and aluminum window frames
Exterior: Interlocking block
Interior finishes
Floor/wall: Porcelain tile
Ceiling: NAD coating

Sanitary fixtures used
(TOTO products)
Toilet: UAXC3CS1/
Washlet (with Eco Remote Controller): TCF5840AUPR/
Urinal: XPU21A/ Wash basin: LS721/ Automatic faucet: TENA12AW/
Ostomate sink: UAS81RSB2NW/
Baby seat: YKA16R

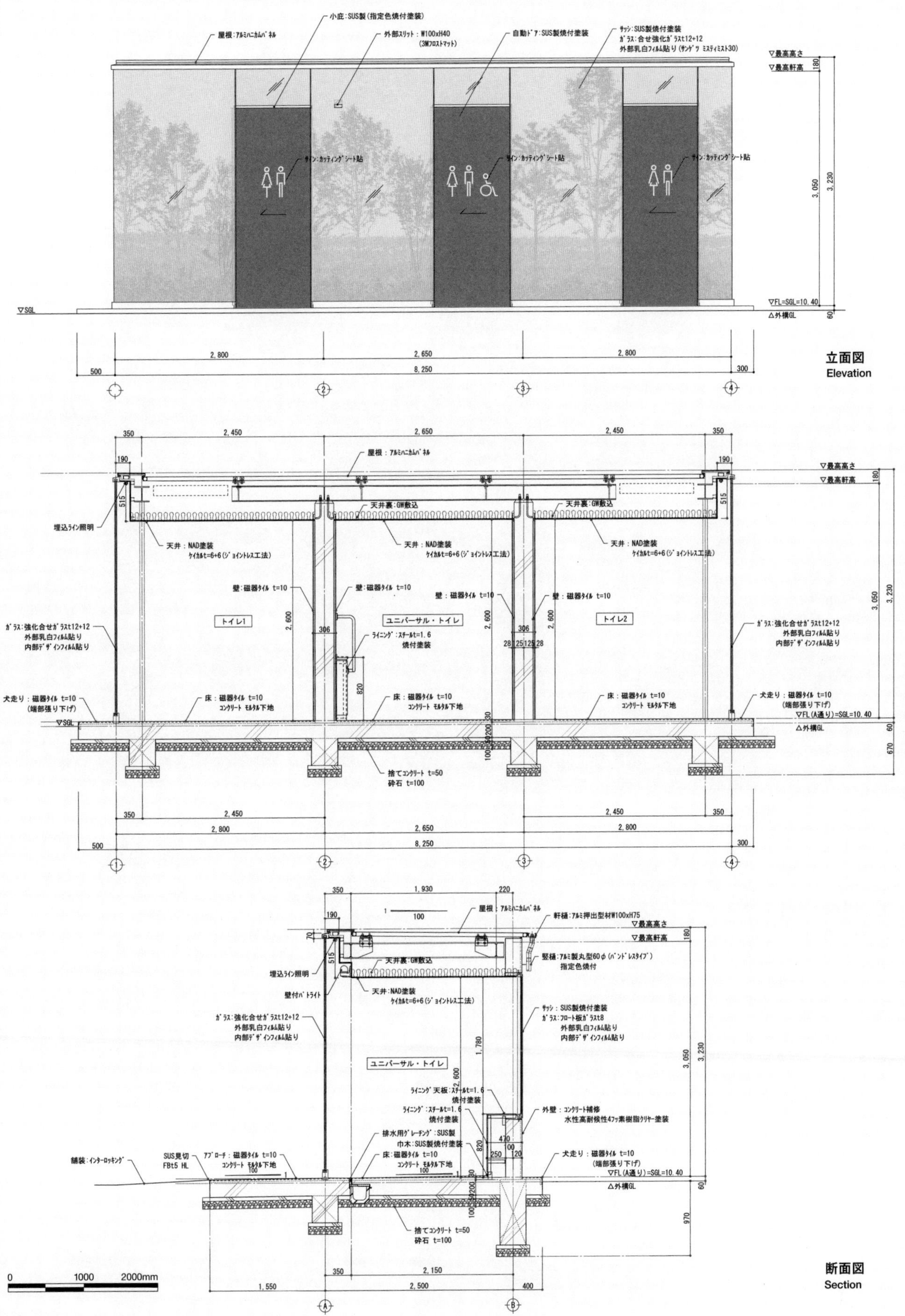

立面図
Elevation

断面図
Section

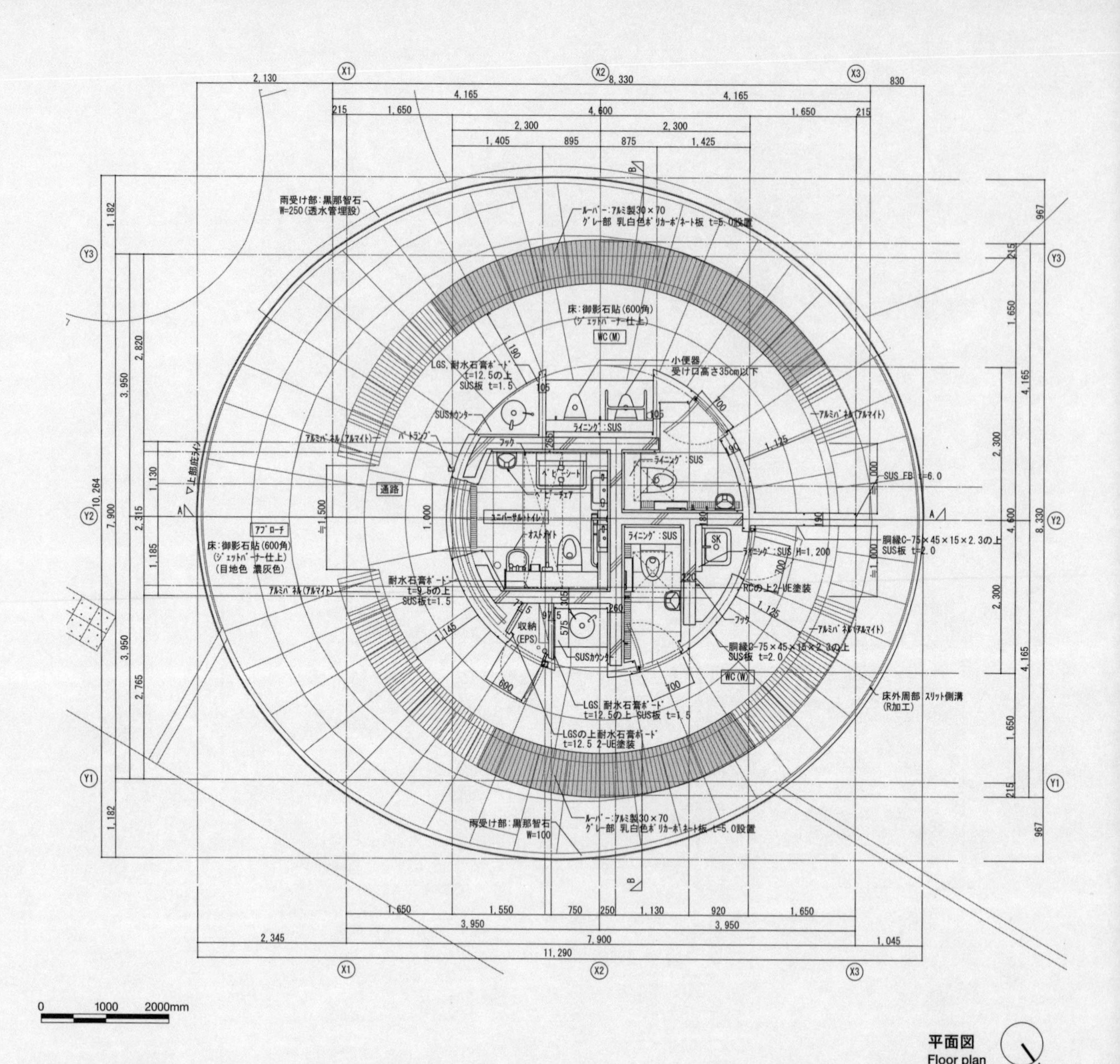

平面図
Floor plan

神宮通公園トイレ P.72

JINGU-DORI PARK PUBLIC TOILET

【所在地】東京都渋谷区神宮前6-22-8
【デザイン】安藤忠雄建築研究所／
安藤忠雄　水谷孝明
【設計・施工】大和ハウス工業

【規模】敷地面積　340.77m²／
建築面積　62.80m²／延床面積　54.47m²
建蔽率　18.42%(許容：建築基準法80%
都市公園法公園全体面積の2%)
【寸法】最高高　3,550mm／
軒高　3,300mm
【敷地条件】商業地域　防火地域
60m高度地区
【構造】鉄筋コンクリート造　一部鉄骨造
【施工期間】2020年3月～9月

【外部仕上げ】
屋根：ルーフデッキプレートの上、デッキコンクリート t=130mm 高強度高伸長ウレタン塗膜防水
庇：アルミハニカム t=104mm 焼付塗装
外壁：アルミパイプ焼付塗装
開口部：ステンレス建具バイブレーション仕上げ
外構：御影石 t=25mm ジェットバーナー仕上げ

【内部仕上げ】
床：御影石 t=25mm ジェットバーナー仕上げ
壁：ステンレスプレート t=1.5mm バイブレーション仕上げ
天井：セメントボード t=8mmの上、2-UE塗装

【主な使用機器】(TOTO)
壁掛大便器：UAXC3CS1／
ウォシュレット(エコリモコン)：TCF5840AUPR
壁掛小便器：XPU21A／
汚垂石：ハイドロセラ・フロアPU：AB690G／
コンパクト多機能トイレパック：
UADAK21L1A1ASN2BA／
台付自動水栓：TENA40AW／ベビーチェア：
YKA15R／ベビーシート：YKA25R

Location
6-22-8 Jingumae, Shibuya-ku, Tokyo
Design
Tadao Ando Architect & Associates/
Tadao Ando, Takaaki Mizutani
Design and construction
Daiwa House Industry

Size
Site area 340.77 m²/
Building area 62.80 m²/
Total floor area 54.47 m²/
Building coverage ratio 18.42%
(Maximum allowable ratio:
Building Standards Act - 80%/
Urban Park Act - 2% of total park area)
Dimension
Maximum building height 3,550 mm/
Eave height 3,300 mm
Site condition
Commercial District,
Fire Prevention District,
60 m-Height Control District
Structure
Reinforced concrete construction,
partially steel construction
Construction period
March - September 2020

Exterior finishes
Roof: Concrete t=130mm on roof deck plates with high-strength/high-stretch polyurethane coating waterproofing
Eaves: Aluminum honeycomb panel t=104 mm with baked coating
Exterior wall: Aluminum pipe with baked coating/ Opening: Stainless steel fittings with vibration finish/ Exterior: Granite t=25mm with flamed finish
Interior finishes
Floor: Granite t=25mm with flamed finish
Wall: Stainless steel plate t=1.5 mm with vibration finish/ Ceiling: Cement board t=8mm with 2-UE coating

Sanitary fixtures used
(TOTO products)
Toilet: UAXC3CS1/
Washlet (with Eco Remote Controller): TCF5840AUPR/
Urinal: XPU21A/ Large ceramic floor tile (for use under urinals): AB690G/
Wheelchair accessible toilet unit: UADAK21L1A1ASN2BA/
Automatic faucet: TENA40AW/
Baby seat: YKA15R/
Baby changing station: YKA25R

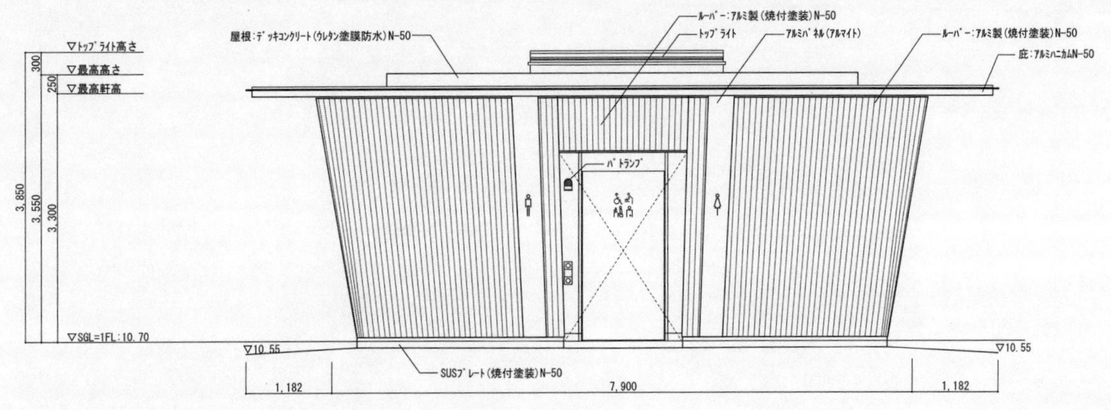

立面図
Elevation

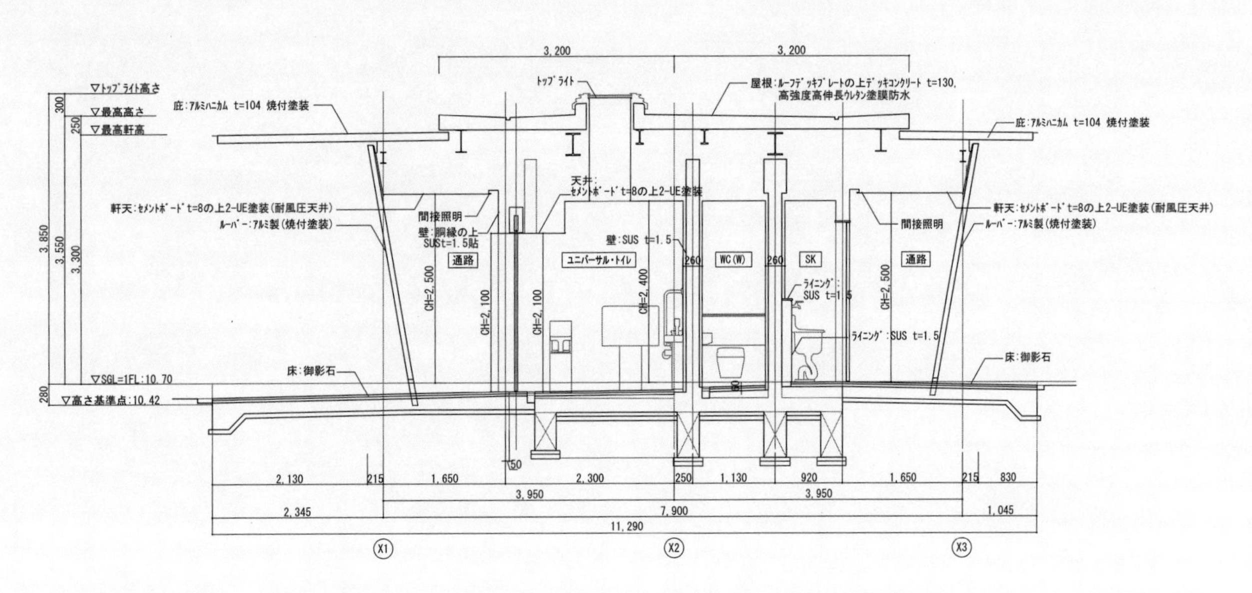

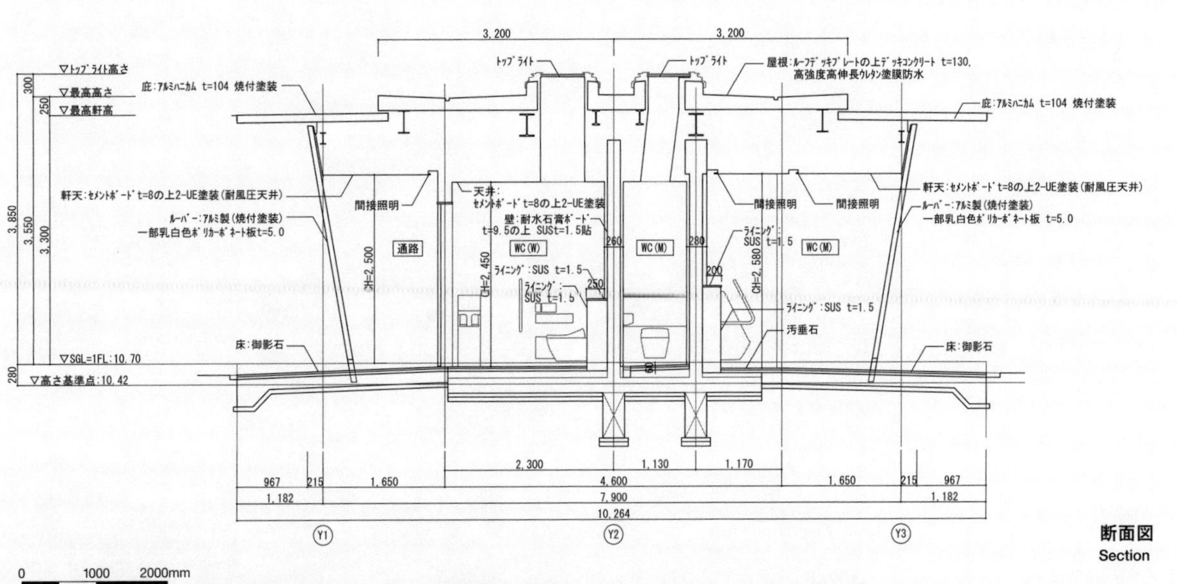

断面図
Section

0 1000 2000mm

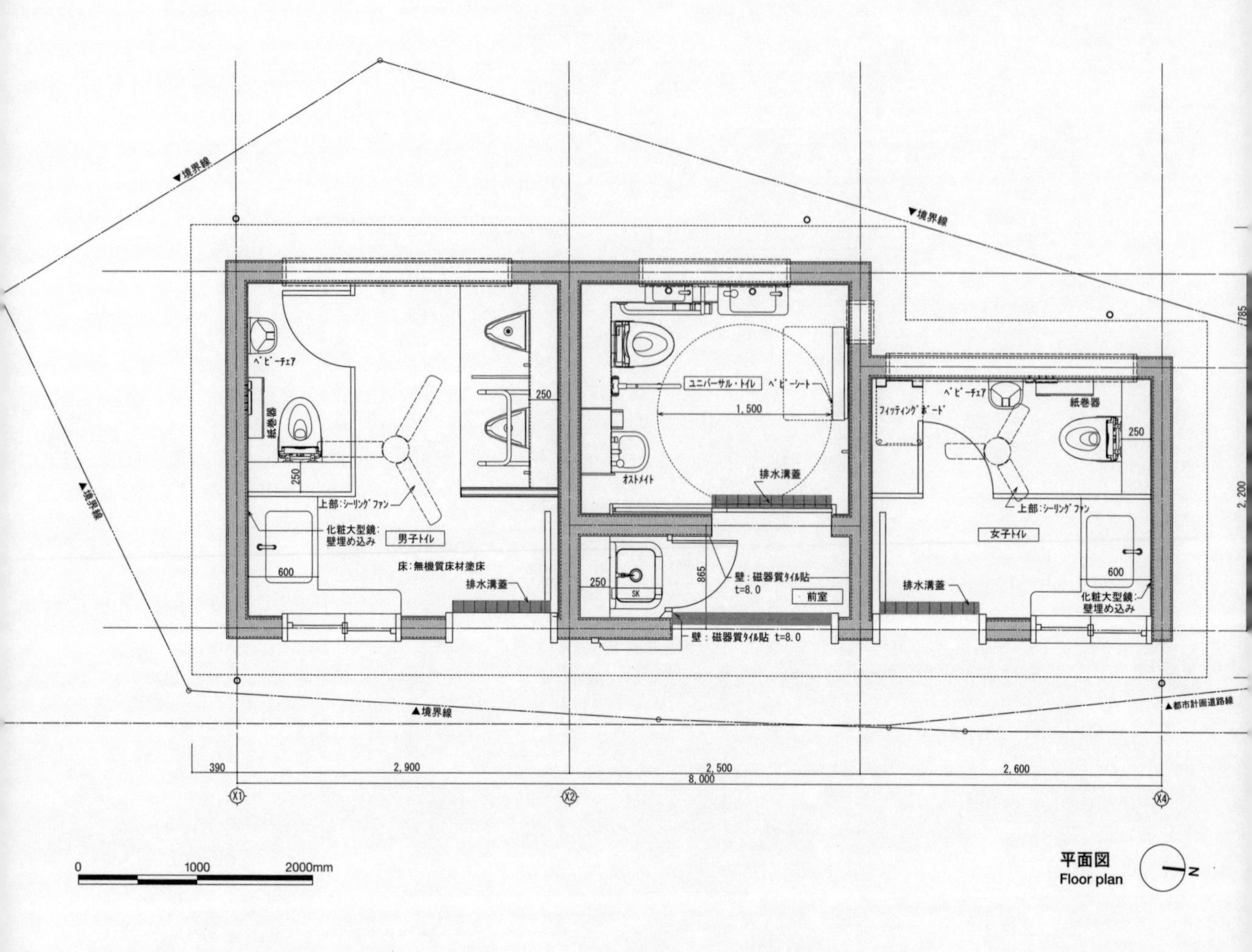

平面図
Floor plan

神宮前公衆トイレ P.82

JINGUMAE PUBLIC TOILET

【所在地】 東京都渋谷区神宮前1-3-14
【デザイン】 NIGO®
【設計・施工】 大和ハウス工業

【規模】 敷地面積　53.93m²(道路占用面積)
建築面積　21.85m²／
延床面積　21.85m²
【寸法】 最高高　4,440mm／
軒高　2,820mm
【敷地条件】 商業地域　防火地域
60m高度地区
【構造】 壁式鉄筋コンクリート造　一部鉄骨造
【施工期間】 2021年2月〜5月

【外部仕上げ】
屋根: 瓦葺き
外壁: コンクリート打ち放しの上、
フッ素塗装ローラー仕上げ
開口部: スチール製アクリル焼付塗装／
ステンレス製焼付塗装
外構: インターロッキング舗装／
アスファルト舗装
【内部仕上げ】
床: エポキシ樹脂系塗床
壁: セラミックタイル
天井: 硬質木片セメント板表し、EP塗装

【主な使用機器】(TOTO)
壁掛大便器: UAXC3CS1／
ウォシュレット(エコリモコン): TCF5840AUPS
壁掛小便器: XPU21A／
ハイドロセラ・フロアPU: AB690G／
コンパクト多機能トイレパック:
UADAK21R1A1ASN2WA／
ベビーチェア: YKA15R／
ベビーシート: YKA25R／
フィッティングボード: YKA41

Location
1-3-14 Jingumae, Shibuya-ku, Tokyo
Design
NIGO®
Design and construction
Daiwa House Industry
Size
Site area 53.93 m² (Road occupancy area)/
Building area 21.85 m²/
Total floor area 21.85 m²
Dimension
Maximum building height 4,440 mm/
Eave height 2,820 mm
Site condition
Commercial District,
Fire Prevention District,
60 m-Height Control District
Structure
Reinforced concrete wall construction,
partially steel construction
Construction period
February - May 2021

Exterior finishes
Roof: Roofing tile
Exterior wall: Exposed concrete with
roller-applied fluorine coating
Opening: Steel with acrylic baked coating
and stainless steel with baked coating
Exterior: Interlocking paving and asphalt
paving
Interior finishes
Floor: Epoxy resin floor coating
Wall: Ceramic tile
Ceiling: Rigid wood chip cement board
with emulsion paint finish

Sanitary fixtures used
(TOTO products)
Toilet: UAXC3CS1/
Washlet (with Eco Remote Controller):
TCF5840AUPS/
Urinal: XPU21A/
Large ceramic floor tile
(for use under urinals): AB690G/
Wheelchair accessible toilet unit:
UADAK21R1A1ASN2WA/
Baby seat: YKA15R/
Baby changing station: YKA25R/
Changing board: YKA41

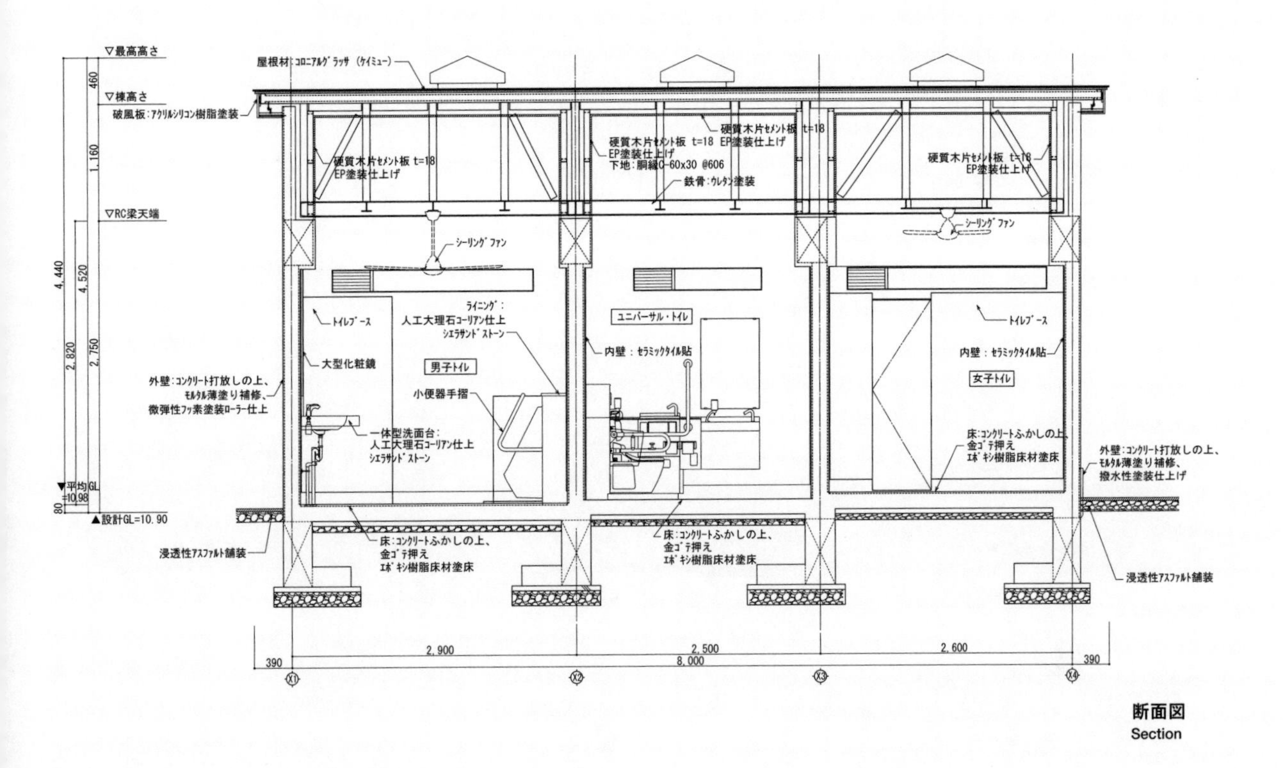

断面図
Section

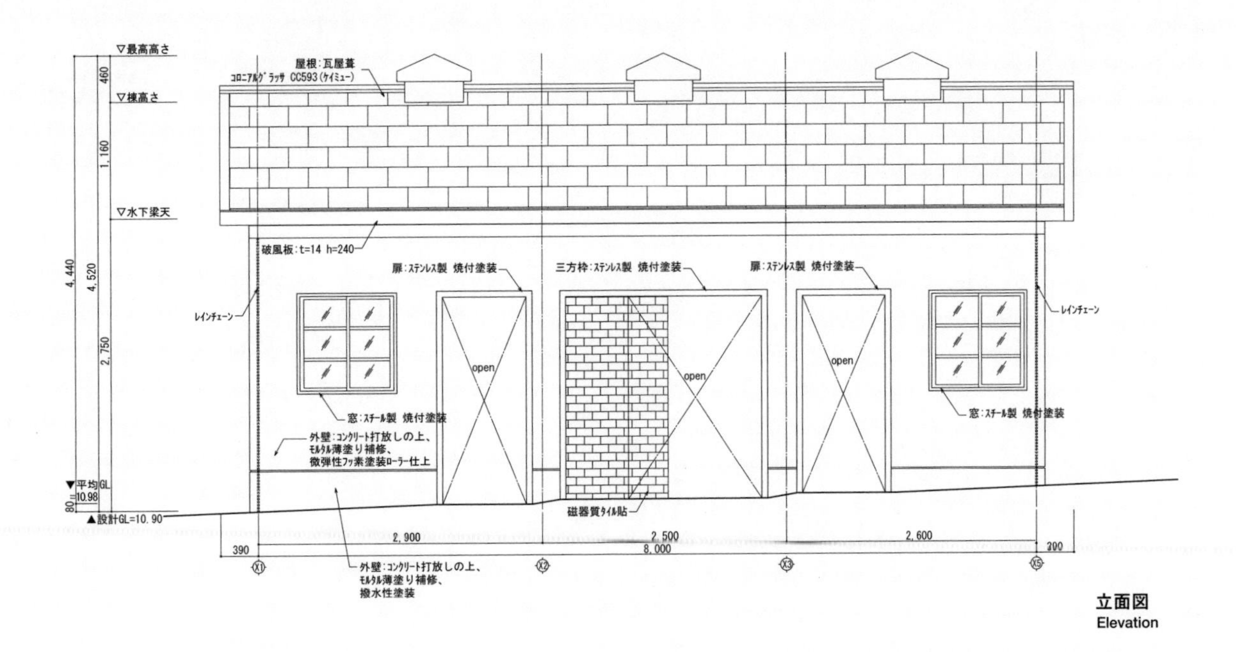

立面図
Elevation

鍋島松濤公園トイレ P.92

NABESHIMA SHOTO PARK PUBLIC TOILET

【所在地】 東京都渋谷区松濤2-10-7
【デザイン】 隈研吾建築都市設計事務所／隈研吾
芳井菜穂子　山縣萌子　キムジョンウォン
菅原史生　間瀬京子
【設計・施工】 大和ハウス工業

【規模】 敷地面積　394.09m²／建築面積　24.41m²
延床面積　24.41m²／建蔽率　6.19%(許容: 60%)
【寸法】 最高高　3,950mm／2,750mm／
3,550mm／2,650mm
軒高　3,650mm／2,450mm／3,250mm／
2,350mm　天井高　2,200mm／2,700mm
【敷地条件】 第一種低層住居専用地域
準防火地域　第一種高度地区
【構造】 鉄筋コンクリート造
【施工期間】 2020年12月〜2021年6月

【外部仕上げ】
屋根: コンクリート金ゴテ仕上の上、ウレタン塗膜防水
外壁: コンクリート　アクリルシリコン樹脂サンドセラミック調装飾仕上塗材　特殊ローラー仕上＋天然木ルーバー防腐剤塗装の上クリア塗装／開口部: スチール製ドアUE塗装／外構: セメント系木質舗装
【内部仕上げ】
床: 磁器質タイル
壁: ケイ酸カルシウム板　EP塗装＋化粧木材
天井: ケイ酸カルシウム板　EP塗装

【主な使用機器】(TOTO)
パブリックコンパクト便器: CFS498BC／ウォシュレット(エコリモコン): TCF5840AUPS／自動洗浄小便器: UFS900JCS／ハイドロセラ・フロアPU:AB680G
壁掛手洗器: LSE90BAPR／手洗器: LS901／台付自動水栓: TENA12BL／コンパクト多機能トイレパック: UADBK61R1A1ASN2WA／幼児用大便器: CS300B+S300BK／幼児用小便器: U310GY／ベビーシート: YKA25R／フィッティングボード: YKA41

Location
2-10-7 Shoto, Shibuya-ku, Tokyo
Design
Kengo Kuma & Associates/ Kengo Kuma, Nahoko Yoshii, Moeko Yamagata, Jeongwon Kim, Fumiki Sugawara, Kyoko Mase
Design and construction
Daiwa House Industry

Size Site area 394.09m²/ Building area 24.41m²/ Total floor area 24.41m²/ Building coverage ratiom²6.19% (Maximum allowable ratio 60%)
Dimension
Maximum building height 3,950 mm/ 2,750 mm/3,550 mm/2,650 mm
Eave height 3,650 mm/2,450 mm/3,250 mm/ 2,350 mm/Ceiling height 2,200 mm/2,700 mm
Site condition
Category 1 Low-rise Exclusive Residential District, Quasi-fire Prevention District, Category 1 Height Control District
Structure Reinforced concrete construction
Construction period
December 2020 - June 2021

Exterior finishes
Roof: Metal trowel-finish concrete with urethane waterproofing coating/ Exterior wall: Concrete with acrylic silicone resin sand-ceramic texture decorative finish coating, special roller finish/ Natural wood louvers, clear-coated over antiseptic coating
Opening: Steel door with UE coating
Exterior: Cement-based wooden paving
Interior finishes Floor: Porcelain tile
Wall: Calcium silicate board with emulsion paint finish/ Decorative wood/ Ceiling: Calcium silicate board with emulsion paint finish

Sanitary fixtures used (TOTO products)
Toilet: CFS498BC/ Washlet (with Eco Remote Controller): TCF5840AUPS/ Urinal: UFS900JCS/ Large ceramic floor tile (for use under urinals): AB680G/ Hand wash basin: LSE90BAPR/ Hand wash basin: LS901/ Automatic faucet: TENA12BL/ Wheelchair accessible toilet unit: UADBK61R1A1ASN2WA/ Infant toilet: CS300B+S300BK/ Infant urinal: U310GY/ Baby changing station: YKA25R/ Changing board: YKA41

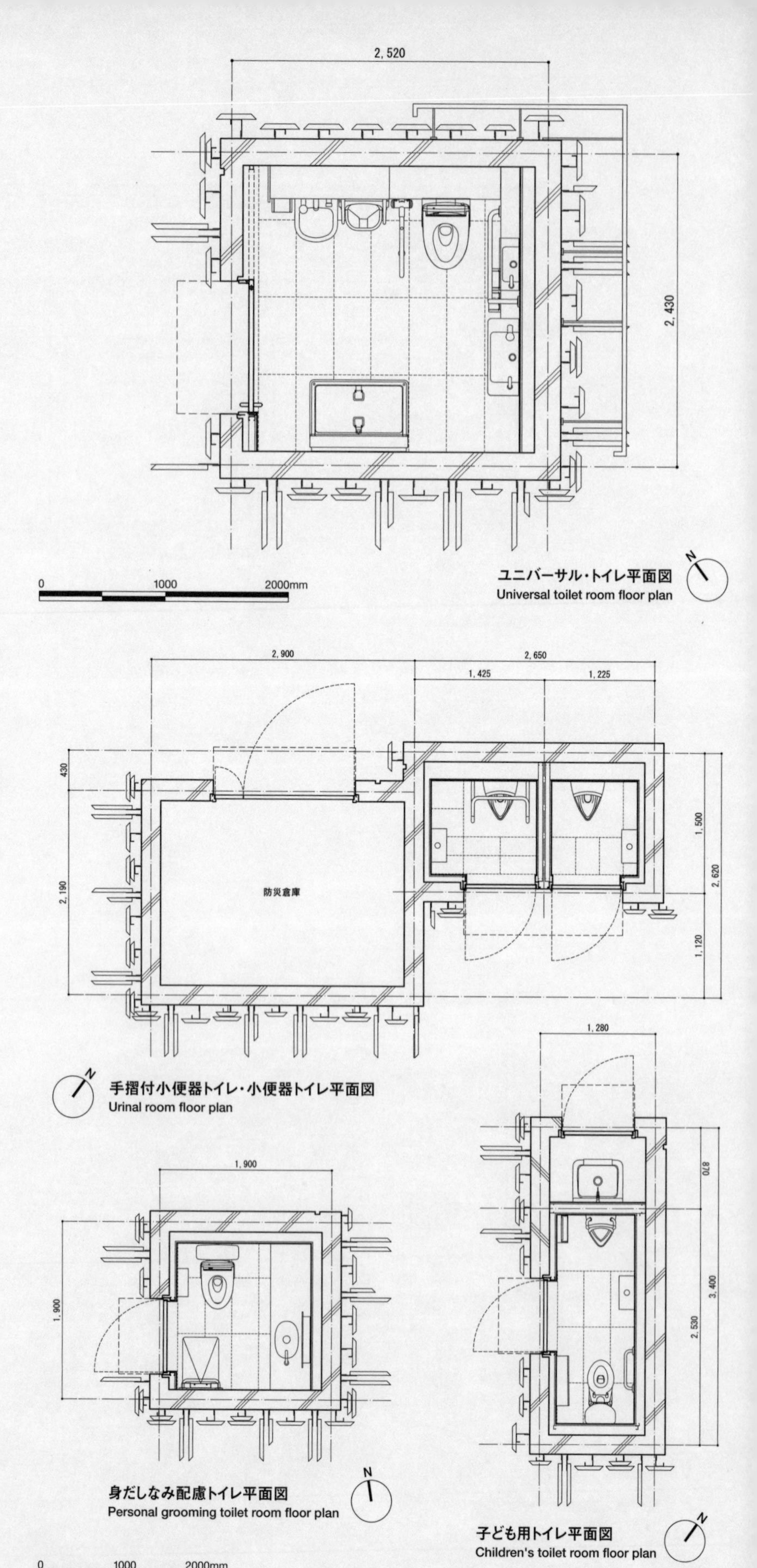

ユニバーサル・トイレ平面図
Universal toilet room floor plan

手摺付小便器トイレ・小便器トイレ平面図
Urinal room floor plan

身だしなみ配慮トイレ平面図
Personal grooming toilet room floor plan

子ども用トイレ平面図
Children's toilet room floor plan

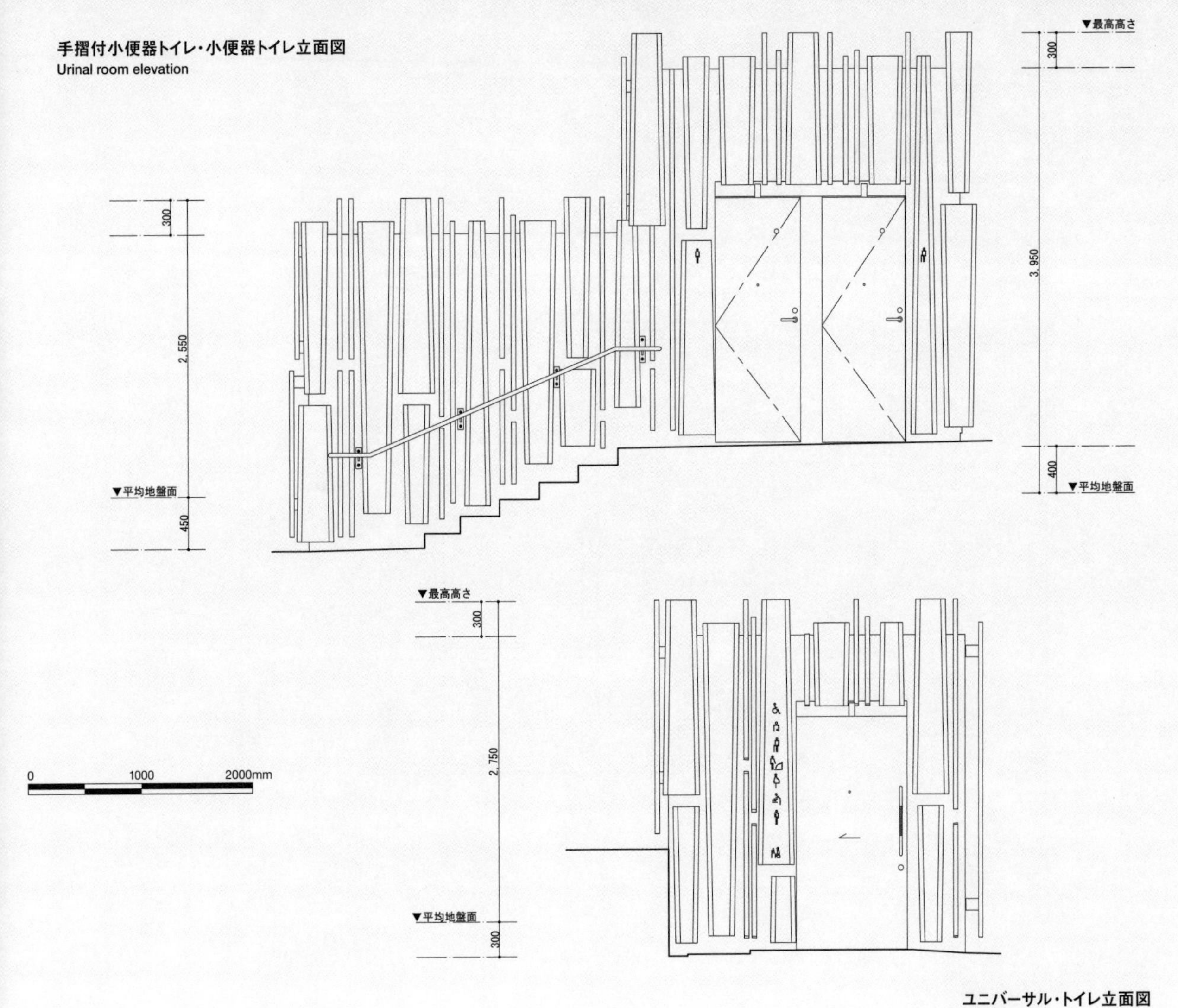

手摺付小便器トイレ・小便器トイレ立面図
Urinal room elevation

ユニバーサル・トイレ立面図
Universal toilet room elevation

▼最高高さ
300
3,550
▼平均地盤面
1,280
▼最高高さ
300
2,750
▼平均地盤面
300
2,430

0 1000 2000mm

子ども用トイレ・ユニバーサル・トイレ 断面図
Section through the children's and men's toilet rooms

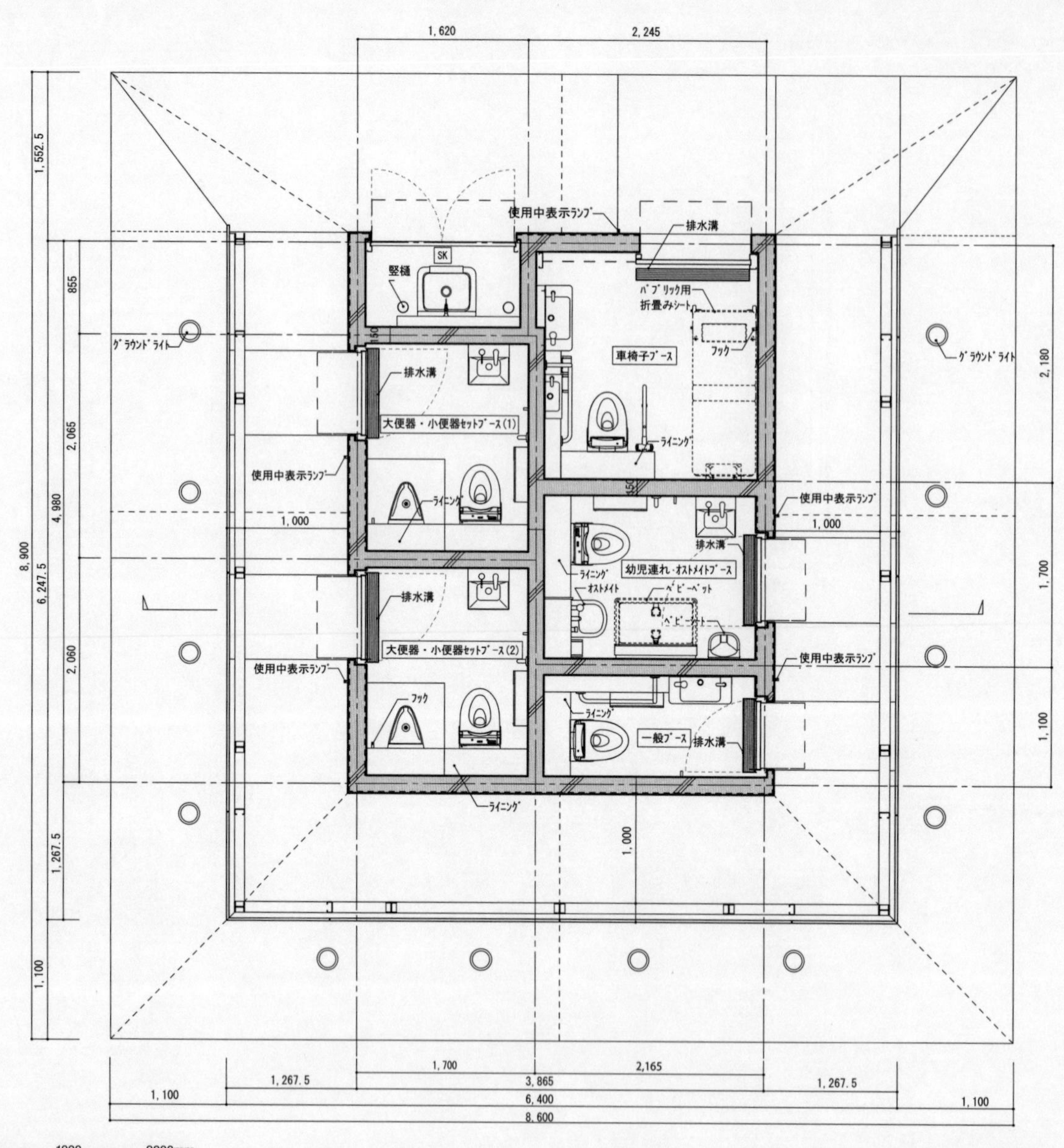

平面図
Floor plan

恵比寿駅西口公衆トイレ P.102

EBISU STATION, WEST EXIT PUBLIC TOILET

【所在地】東京都渋谷区恵比寿南1-5-8
【デザイン】サムライ／佐藤可士和
齊藤良博　石川耕
【設計・施工】大和ハウス工業

【規模】敷地面積　道路上建築のため
敷地設定なし／建築面積　19.25m²／
延床面積　19.25m²
【寸法】最高高　3,790mm／
軒高　3,280mm／天井高　2,400mm
【敷地条件】商業地域　防火地域
60m高度地区
【構造】鉄筋コンクリート造　一部鉄骨造
【施工期間】2021年2月～7月

【外部仕上げ】
屋根: ウレタン塗膜防水t=3mm
外壁: コンクリート打ち放し金ゴテ補修の上、
低汚染型水性無機質塗装
+落書き防止クリア塗料
開口部: スチール製建具焼付塗装
（ウレタン塗装上塗り）
外部: アルミルーバー フッ素焼付塗装
外構: エポキシ樹脂系塗床材(防滑仕様)
【内部仕上げ】
床: エポキシ樹脂系塗床材(防滑仕様)
壁: コンクリート打ち放し金ゴテ補修の上、低汚染型水性無機質塗装+落書き防止クリア塗料
天井: ケイカル板 t=6mmの上、EP塗装

【主な使用機器】(TOTO)
壁掛大便器: UAXC3CS1／
ウォシュレット（エコリモコン）: TCF5840AUPS
壁掛小便器: XPU21A／
ハイドロセラ・フロアPU: AB680BR／
洗面器: LS721CM／台付自動水栓:
TENA40AW／水石けん入れ: TLK05202J
コンパクト多機能トイレパック:
UADAK01R1A1ANN2WA／
コンパクトオストメイトパック: UAS81RSB2NW
パブリック用折りたたみシート: EWC500RR
ベビーチェア: YKA15R／ベビーシート: YKA25R

Location
1-5-8 Ebisuminami, Shibuya-ku, Tokyo
Design
SAMURAI/ Kashiwa Sato, Yoshihiro Saito, Ko Ishikawa
Design and construction
Daiwa House Industry

Size
Site area - N/A (Non-site specific structure constructed on road)/
Building area 19.25m²/
Total floor area 19.25m²
Dimension
Maximum building height 3,790 mm/
Eave height 3,280 mm/
Ceiling height 2,400 mm
Site condition
Commercial District, Fire Prevention District, 60 m-Height Control District
Structure
Reinforced concrete construction, partially steel construction

Construction period
February - July 2021

Exterior finishes
Roof: Urethane waterproof coating t=3 mm/ Exterior wall: Metal-trowel-finished exposed concrete with Low-contamination water-based inorganic coating and anti-graffiti clear coating
Opening: Steel fittings with baked coating (urethane topcoat)
Outer envelope: Aluminum louvers with baked fluorine coating
Exterior: Epoxy resin floor coating (anti-slip finish)
Interior finishes
Floor: Epoxy resin floor coating (anti-slip finish)
Wall: Metal-trowel-finished exposed concrete with Low-contamination water-based inorganic coating and anti-graffiti clear coating/ Ceiling: Calcium silicate board t=6mm with emulsion paint finish

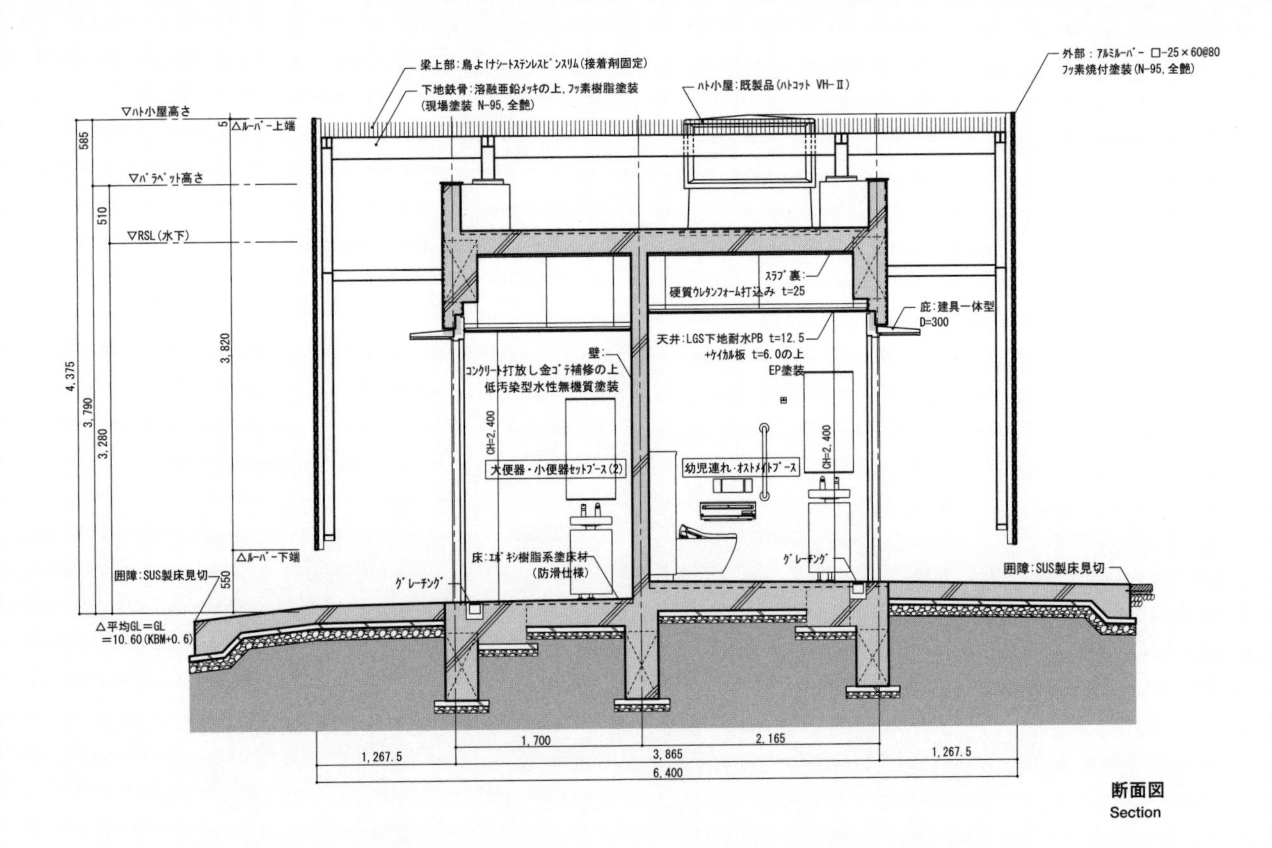

断面図
Section

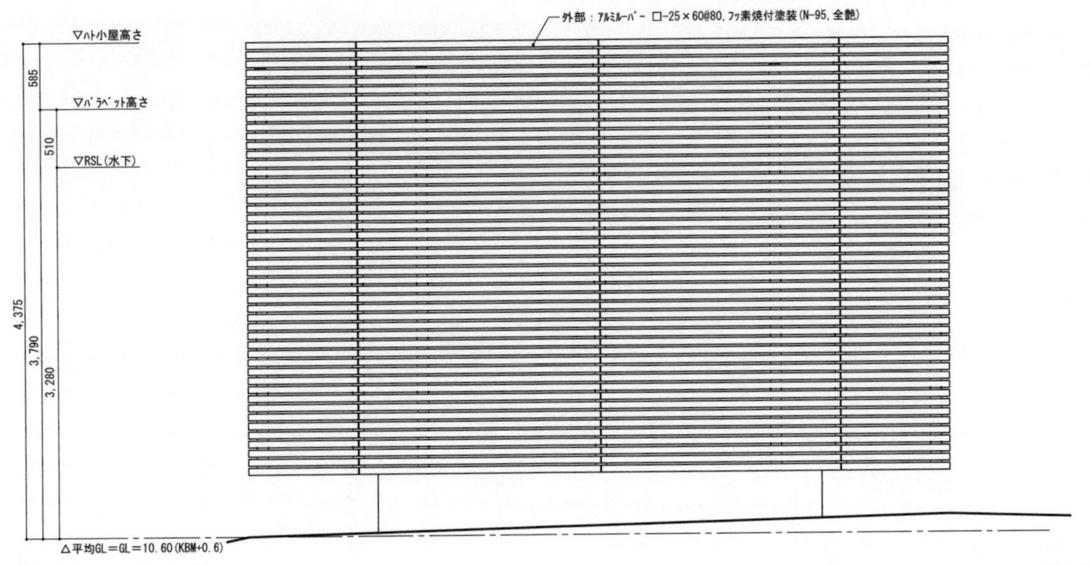

立面図
Elevation

Sanitary fixtures use
(TOTO products)
Toilet: UAXC3CS1/
Washlet (with Eco Remote Controller):
TCF5840AUPS/ Urinal: XPU21A/
Large ceramic floor tile
(for use under urials): AB680BR/
Wash basin: LS721CM/
Automatic faucet: TENA40AW/
Liquid soap dispenser: TLK05202J/
Wheelchair accessible toilet unit:
UADAK01R1A1ANN2WA/ Ostomate sink:
UAS81RSB2NW/ Foldable changing bed:
EWC500RR/ Baby seat:
YKA15R/ Baby changing station:YKA25R

0 1000 2000mm

代々木八幡公衆トイレ P.112

YOYOGI-HACHIMAN PUBLIC TOILET

【所在地】東京都渋谷区代々木5-1-2
【デザイン】伊東豊雄建築設計事務所／伊東豊雄
矢部倫太郎　桝永絵理子(元所員)
【設計・施工】大和ハウス工業
【規模】敷地面積　291.18m²／
建築面積　17.46m²／延床面積　17.46m²／
建蔽率　5.99%(許容: 60%)
【寸法】最高高　3,431mm／軒高　2,931mm／
階高　2,850mm(ユニバーサル・トイレ)／
天井高　3,422mm(ユニバーサル・トイレ)
【敷地条件】準住居地域　準防火地域　第二種高度地区
【構造】鉄筋コンクリート造
【施工期間】2020年10月〜2021年7月

【外部仕上げ】
屋根: ステンレスt=4mmの上遮熱遮音性塗装材
外壁: モザイクタイル貼　光触媒コート
開口部: ステンレスFB　曲げ加工
外壁・建具: モザイクタイル貼
光触媒コート・ステンレスサッシ焼付塗装
外構: モザイクタイル貼
【内部仕上げ】
床: RC　金鏝押えの上浸透性着色表面強化剤塗布
壁: EP　塗装の上落書・貼紙防止コート塗布
天井: 現場発泡ウレタン吹付の上　リシン吹付

【主な使用機器】(TOTO)
壁掛大便器: UAXC3CS1／
ウォシュレット(エコリモコン): TCF5840AUPS／
壁掛小便器: XPU21A／
ハイドロセラ・フロアPU: AB690BR／手洗器: LS901／
台付自動水栓: TLE26502J／
コンパクトオストメイトパック: UAS81RSB2NW

Location
5-1-2 Yoyogi, Shibuya-ku, Tokyo
Design
Toyo Ito & Associates, Architects/
Toyo Ito, Rintaro Yabe,
Eriko Masunaga (Former staff member)
Design and construction
Daiwa House Industry

Size
Site area 291.18 m²/ Building area 17.46 m²/
Total floor area 17.46 m²/
Building coverage ratio 5.99%
(Maximum allowable ratio: 60%)
Dimension
Maximum building height 3,431 mm/
Eave height 2,931 mm/ F
loor height 2,850 mm (universal toilet) /
Ceiling height 3,422 mm (universal toilet)
Site condition
Quasi-residential District, Quasi-fire Prevention District, Category 2 Height Control District
Structure
Reinforced concrete construction
Construction period
October 2020 - July 2021

Exterior finishes
Roof: Stainless steel plate t=4 mm
with thermal and acoustic insulation coating
Exterior wall: mosaic tile with photocatalytic coating
Opening: Bent stainless steel flat bar
Exterior wall/fittings: mosaic tile with photocatalytic coating and stainless steel frames with baked coating
Exterior: mosaic tile
Interior finishes
Floor: Metal trowel-finish concrete with permeable colored surface hardening coating
Wall: Anti-graffiti /flyer pasting coating over emulsion paint finish
Ceiling: Lysine-spraying over in-situ urethane foam spraying

Sanitary fixtures used
(TOTO products)
Toilet: UAXC3CS1/
Washlet (with Eco Remote Controller):
CF5840AUPS/ Urinal: XPU21A/
Large ceramic tile (for use under urinals): AB690BR/
Hand wash basin: LS901/
Automatic faucet: TLE26502J/
Ostomate sink: UAS81RSB2NW

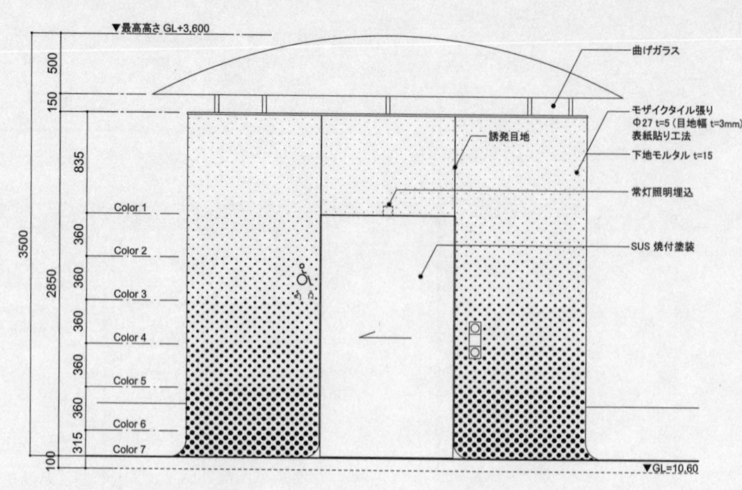

ユニバーサル・トイレ 立面図
Universal toilet room elevation

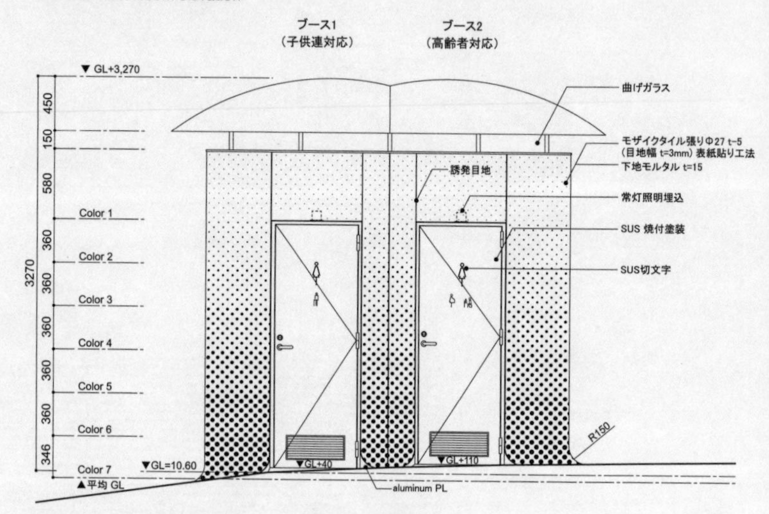

女性用トイレ 立面図
Women's toilet room elevation

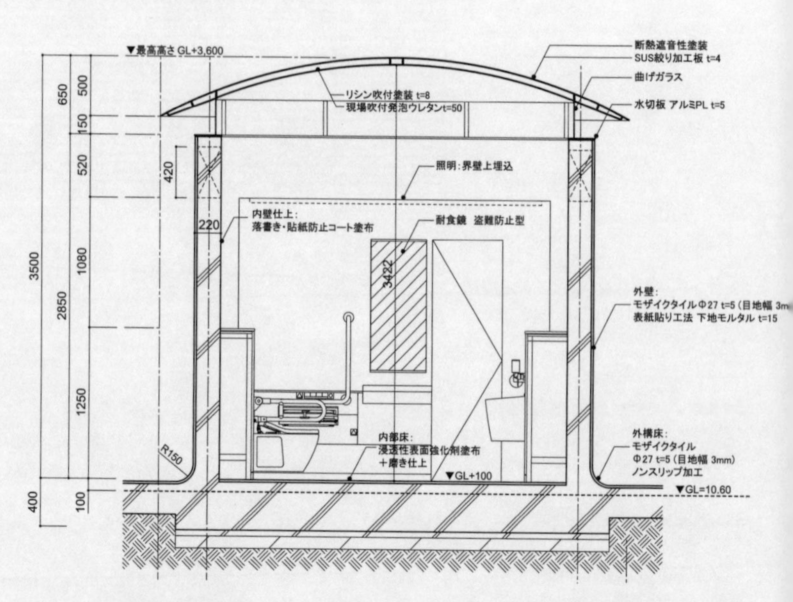

ユニバーサル・トイレ 断面図
Universal toilet room section

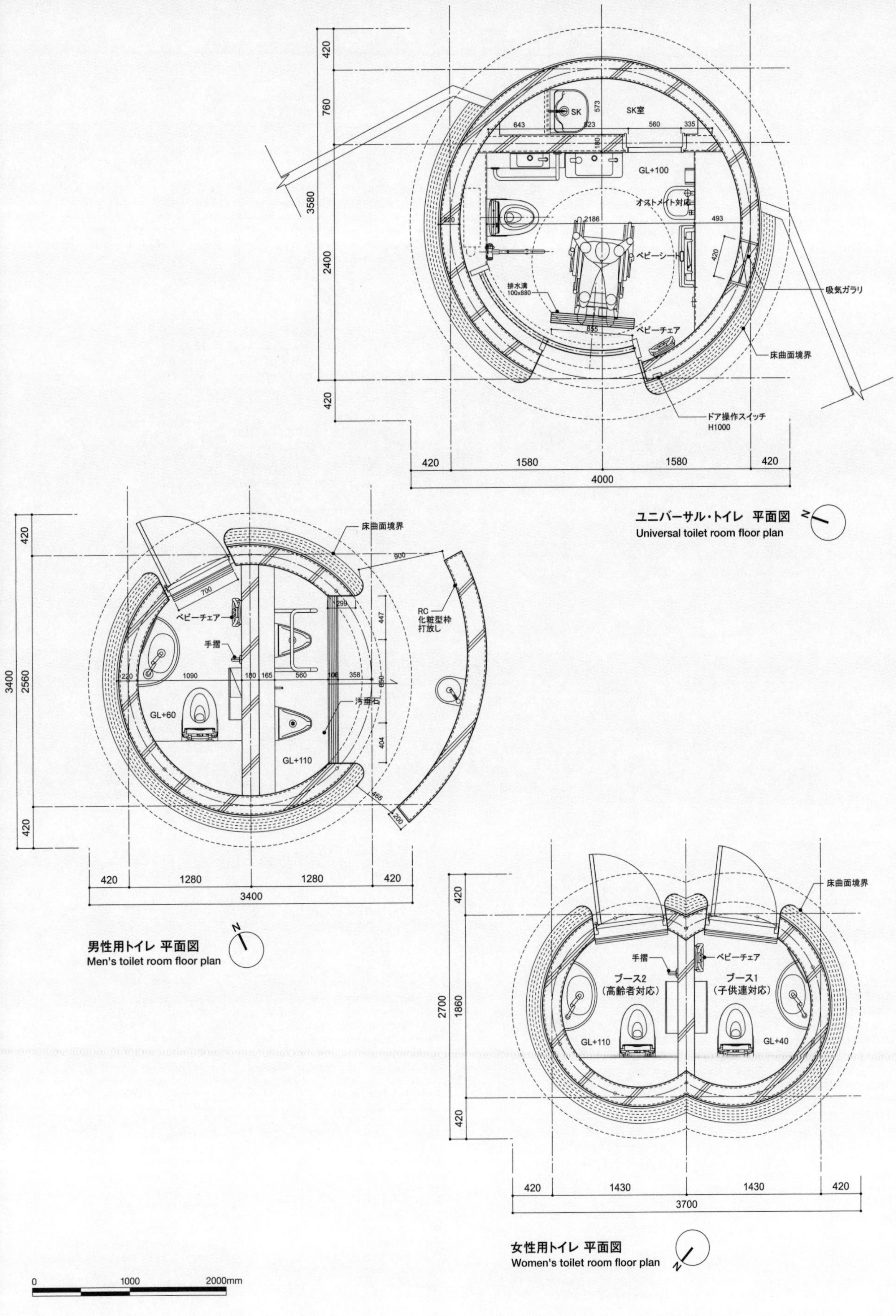

ユニバーサル・トイレ 平面図
Universal toilet room floor plan

男性用トイレ 平面図
Men's toilet room floor plan

女性用トイレ 平面図
Women's toilet room floor plan

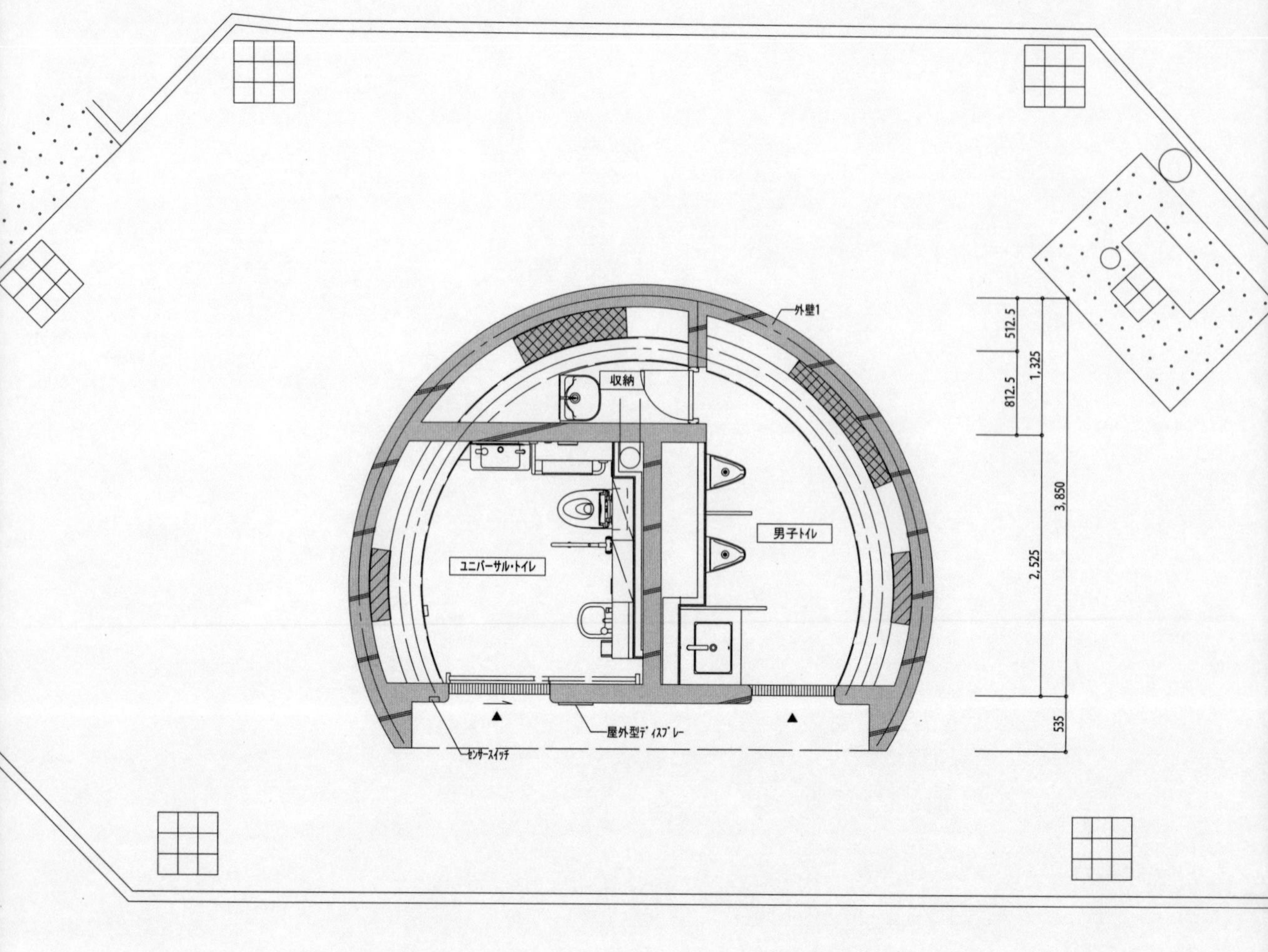

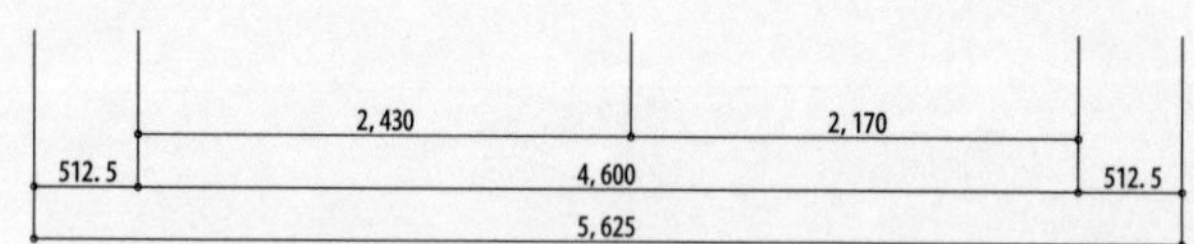

平面図
Floor plan

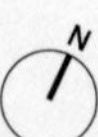

0 1000 2000mm

七号通り公園トイレ P.122
NANAGO DORI PARK PUBLIC TOILET

【所在地】東京都渋谷区幡ヶ谷2-53-5
【デザイン】TBWA\HAKUHODO Disruption Lab Team／佐藤カズー　梅田哲夫　伊藤裕平　畑尾佐助　戸矢 渚
【設計デザイン協力】久保都島建築設計事務所／久保秀朗　都島有美　三浦みづき
【外観デザイン協力】quantum
【ボイスコマンド協力】Birdman
【設計・施工】大和ハウス工業

【規模】敷地面積　195.91m²／建築面積　18.11m²／延床面積　18.11m²
建蔽率　9.24%（許容: 60%）
【寸法】最高高　4,255mm／軒高　4,255mm
【敷地条件】準工業地域　準防火地域
第三種高度地区
【構造】鉄筋コンクリート造
【施工期間】2021年2月～8月

【外部仕上げ】
屋根・外壁: コンクリート下地　高速硬化ウレタン防水＋光触媒塗装
開口部: スチール製建具　ウレタン塗装
外構: カラーアスファルト舗装
【内部仕上げ】
床: ウレタン塗装　防滑仕上
壁: ウレタン塗装
天井: ジョリパット塗装

【主な使用機器】（TOTO）
壁掛大便器: UAXC3CS1／
ウォシュレット（エコリモコン）: TCF5840AUPS
壁掛小便器: XPU21A／洗面器: LS716／
台付自動水栓: TENA12ALW／
コンパクトオストメイトパック: UAS81LSB2NW

Location
2-53-5 Hatagaya, Shibuya-ku, Tokyo
Design
TBWA\HAKUHODO Disruption Lab Team/ Kazoo Sato, Tetsuo Umeda, Yuhei Ito, Sasuke Hatao, Nagisa Toya
Architectural design cooperation
Kubo Tsushima Architects/ Hideaki Kubo, Yumi Tsushima, Mizuki Miura
Exterior design cooperation
quantum
Voice command design cooperation
Birdman
Design and construction
Daiwa House Industry

Size Site area 195.91m²/ Building area 18.11m²/ Total floor area 18.11m²/ Building coverage ratio 9.24% (Maximum allowable ratio 60%)
Dimension
Maximum building height 4,255 mm/ Eave height 4,255 mm
Site condition
Quasi-industrial District, Quasi-fire Prevention District, Category 3 Height Control District
Structure
Reinforced concrete construction

Construction period
February - August 2021

Exterior finishes
Roof/ exterior wall: Concrete substrate with fast curing urethane waterproofing and photocatalytic coating
Opening: Steel fittings with urethane coating
Exterior: Colored asphalt paving
Interior finishes
Floor: Urethane coating (anti-slip finish)
Wall: Urethane coating
Ceiling: Jolypate coating

Sanitary fixtures used
(TOTO products)
Toilet: UAXC3CS1/
Washlet (with Eco Remote Controller): TCF5840AUPS/ Urinal: XPU21A/
Wash basin: LS716/
Automatic faucet: TENA12ALW/
Ostomate sink: UAS81LSB2NW

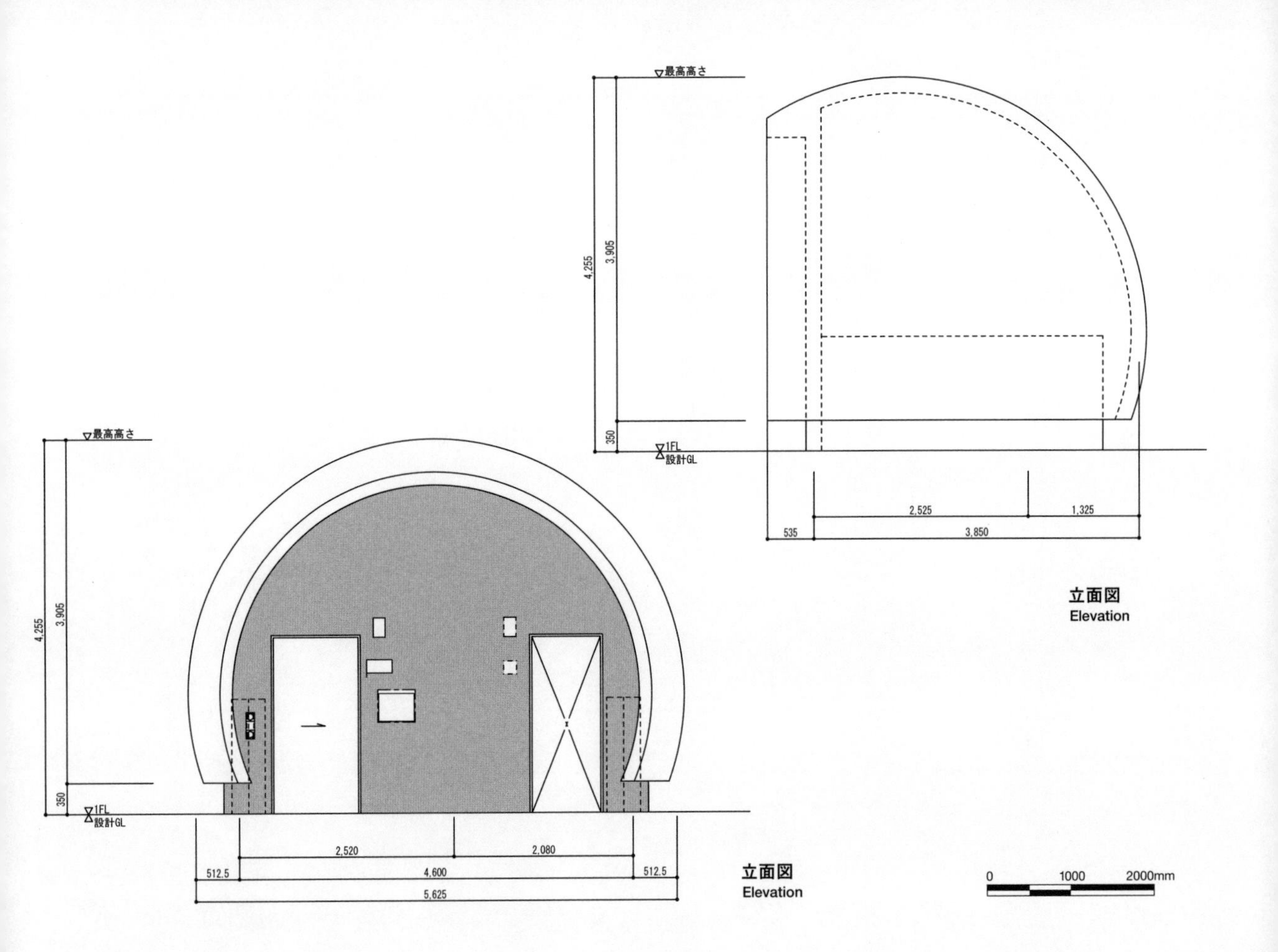

立面図
Elevation

立面図
Elevation

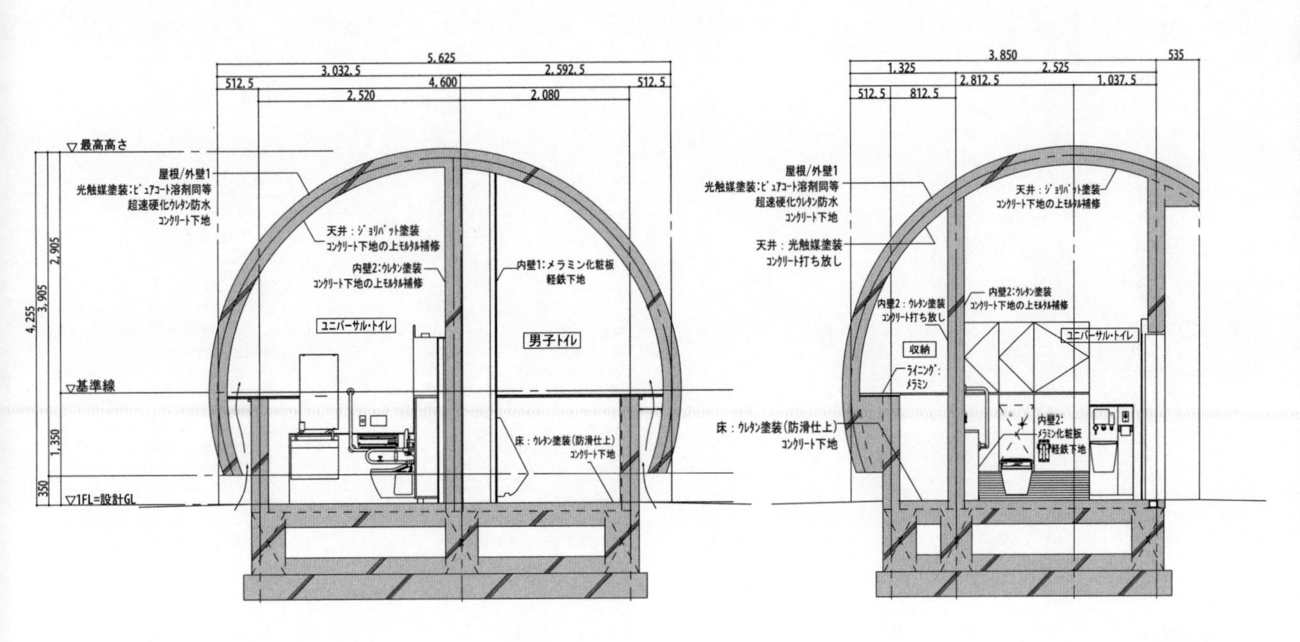

断面図
Section

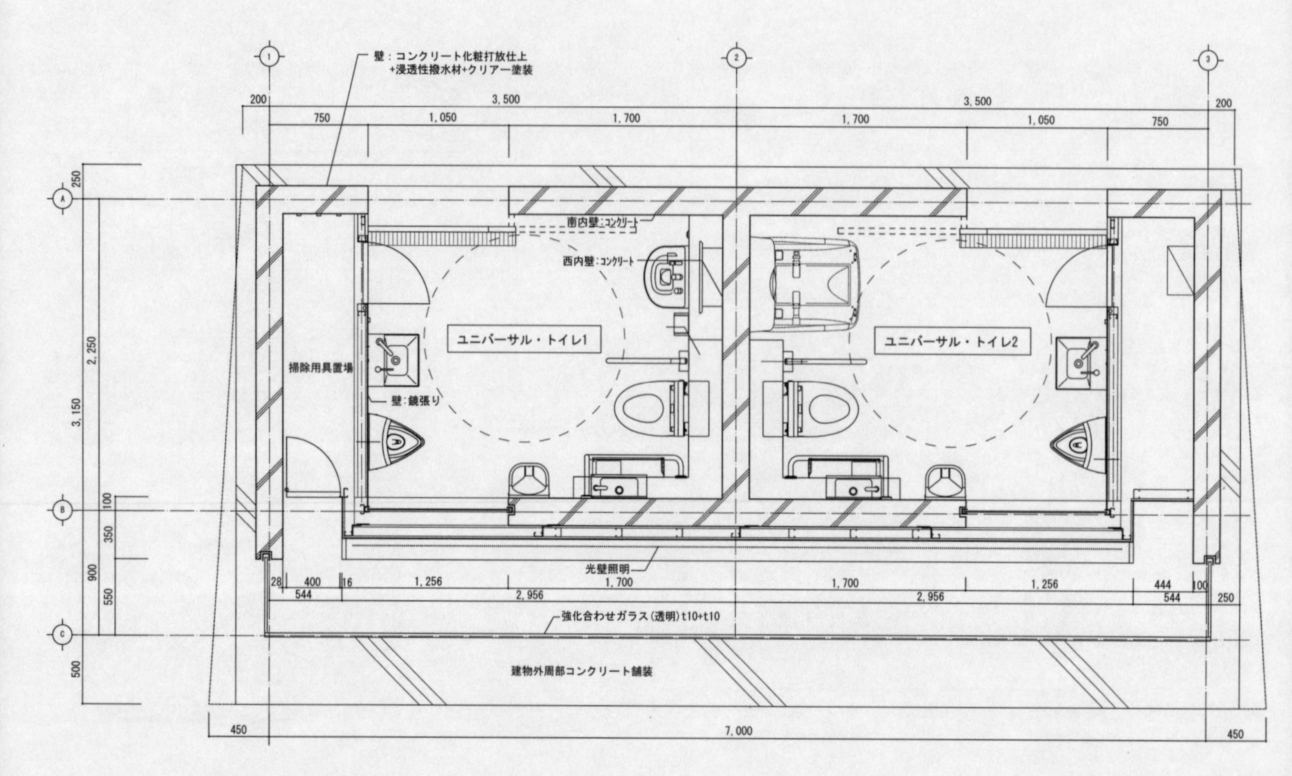

平面図
Floor plan

広尾東公園トイレ P.132

HIROO HIGASHI PARK PUBLIC TOILET

【所在地】東京都渋谷区広尾4-2-27
【デザイン】後智仁　坂下加代子　今城瞬
【設計・施工】大和ハウス工業

【規模】敷地面積　156.71m²／
建築面積　22.12m²／
延床面積　22.12m²／
建蔽率　14.11%（許容：60%）
【寸法】最高高　3,150mm／
軒高　2,865mm／天井高　2,400mm
【敷地条件】第二種中高層住居専用地域・
準防火地域　第三種高度地区
【構造】壁式鉄筋コンクリート造　一部鉄骨造
【施工期間】2022年2月〜7月

【外部仕上げ】
屋根：一般構造用鋼板t=16mm
超耐久フッ素樹脂系遮熱塗料
外壁：コンクリート化粧打ち放し仕上げ+
浸透性撥水材塗布+クリアー塗装
開口部：スチールドア
外構：インターロッキング
【内部仕上げ】
床：アクリル樹脂防塵塗装
壁：浸透性撥水材塗布+クリアー塗装
天井：ケイカル板EP塗装

【主な使用機器】（TOTO）
壁掛大便器：UAXC3CS1／ウォシュレット（エコリモコン）：TCF5840AUPN／壁掛小便器：XPU21A／壁掛洗面器：L721CM／台付自動水栓：TENA40A／オストメイト対応マルチパック：XPSA72C71WW／ベビーシート：YKA24S／ベビーチェア：YKA15S

Location
4-2-27 Hiroo, Shibuya-ku,
Design
Tomohito Ushiro,
Kayoko Sakashita, Shun Imajo
Design and construction
Daiwa House Industry

Size Site area 156.71 m²/
Building area 22.12 m²/
Total floor area 22.12 m²/
Building coverage ratio 14.11%
(Maximum allowable ratio: 60%)
Dimension
Maximum building height 3,150 mm/
Eave height 2,865 mm/
Ceiling height 2,400 mm
Site condition
Category 2 Medium-to-high-rise
Exclusive Residential District,
Quasi-fire Prevention District,
Category 3 Height Control District
Structure
Reinforced concrete wall construction,
partially steel construction
Construction period
February - July 2022

Exterior finishes
Roof: Steel plate for general construction purposes t=16 mm with super-durable fluoropolymer thermal insulation coating
Exterior wall: Exposed fair-faced concrete finish with permeable water-repellent coating and clear coating
Opening: Steel door
Exterior: Interlocking paving
Interior finishes
Floor: Acrylic resin dustproof coating
Wall: Permeable water-repellent coating and clear coating
Ceiling: Calcium silicate board with emulsion paint finish

Sanitary fixtures used
(TOTO products)
Toilet: UAXC3CS1/
Washlet (with Eco Remote Controller): TCF5840AUPN/ Urinal: XPU21A/
Wash basin: L721CM/
Automatic faucet: TENA40A/
Ostomate sink: XPSA72C71:W/
Baby changing station: YKA24S/
Baby seat: YKA15S

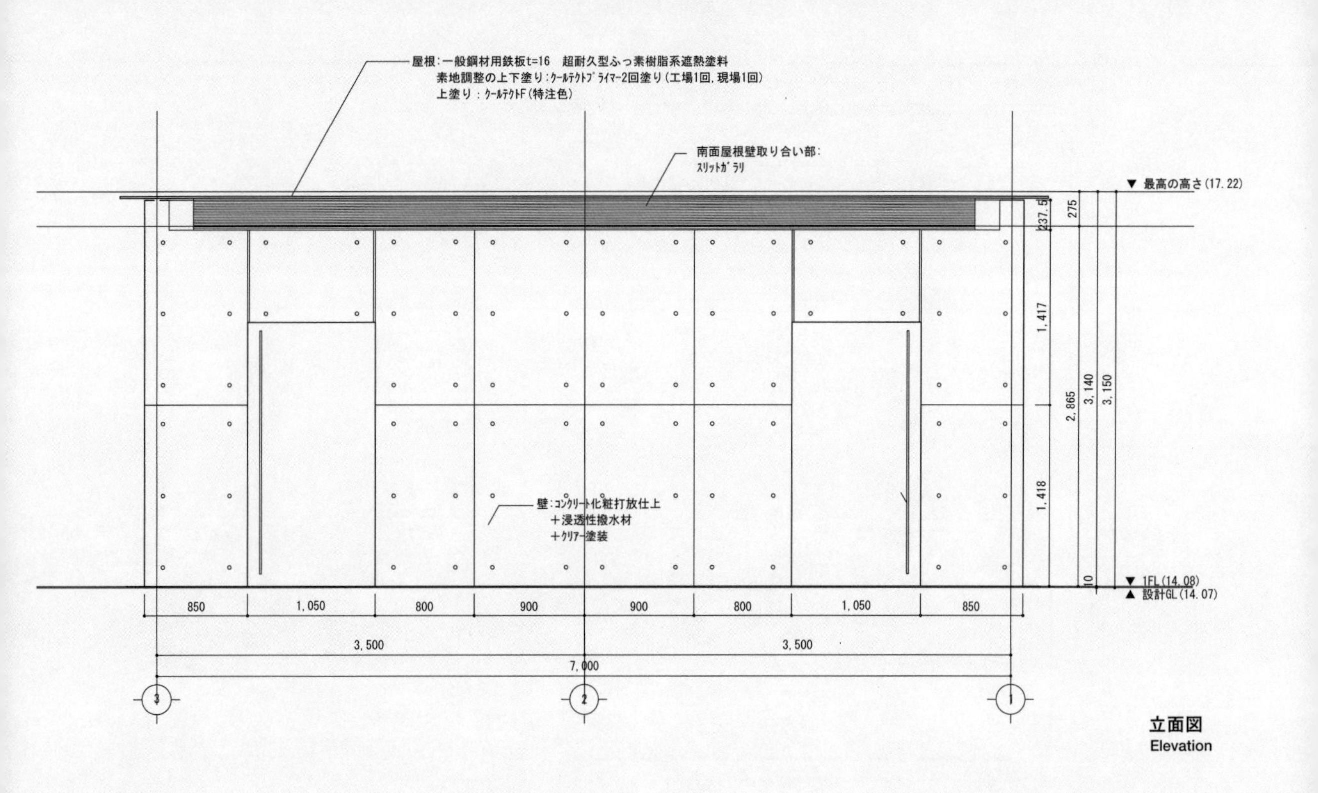

立面図
Elevation

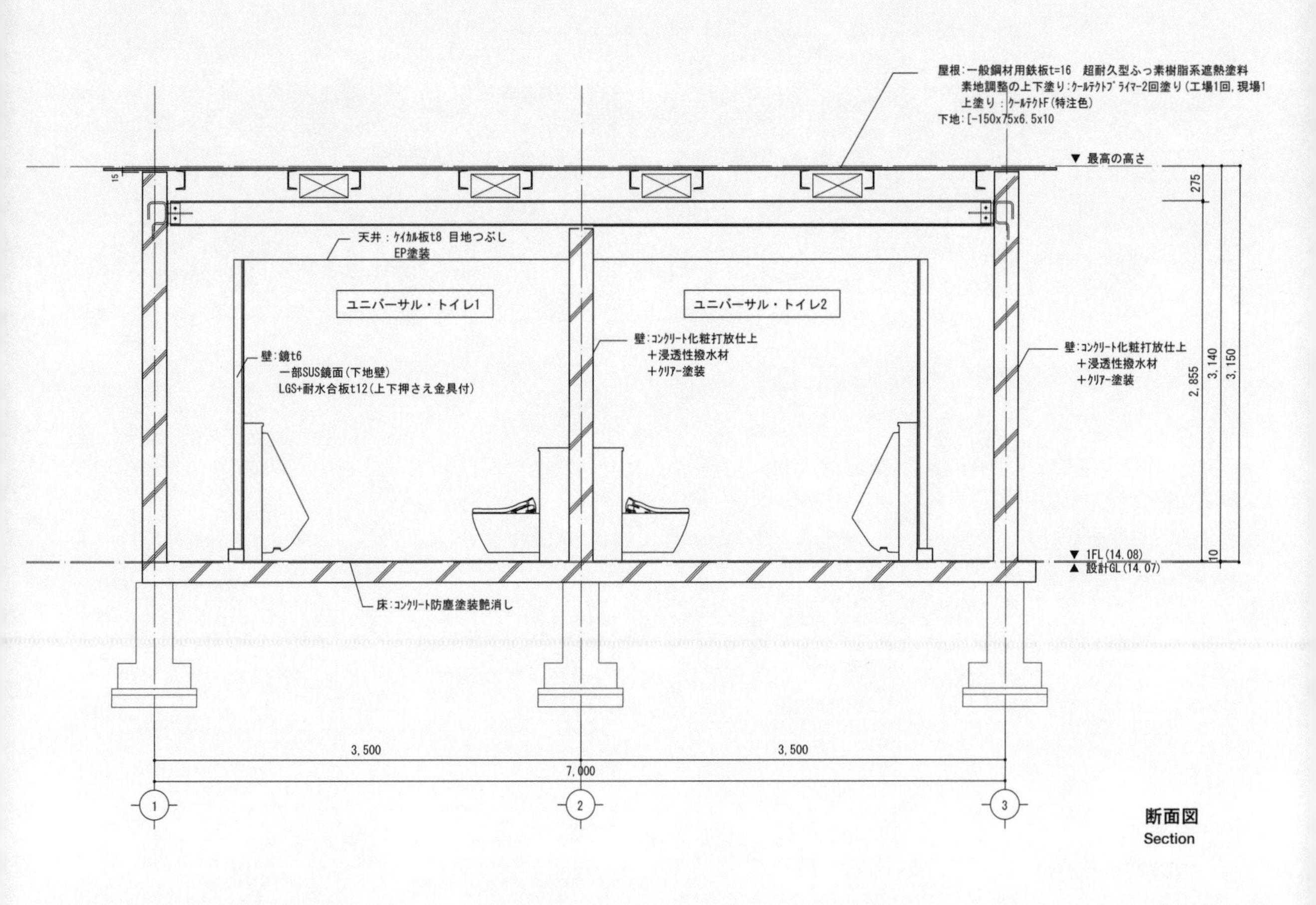

断面図
Section

0　1000　2000mm

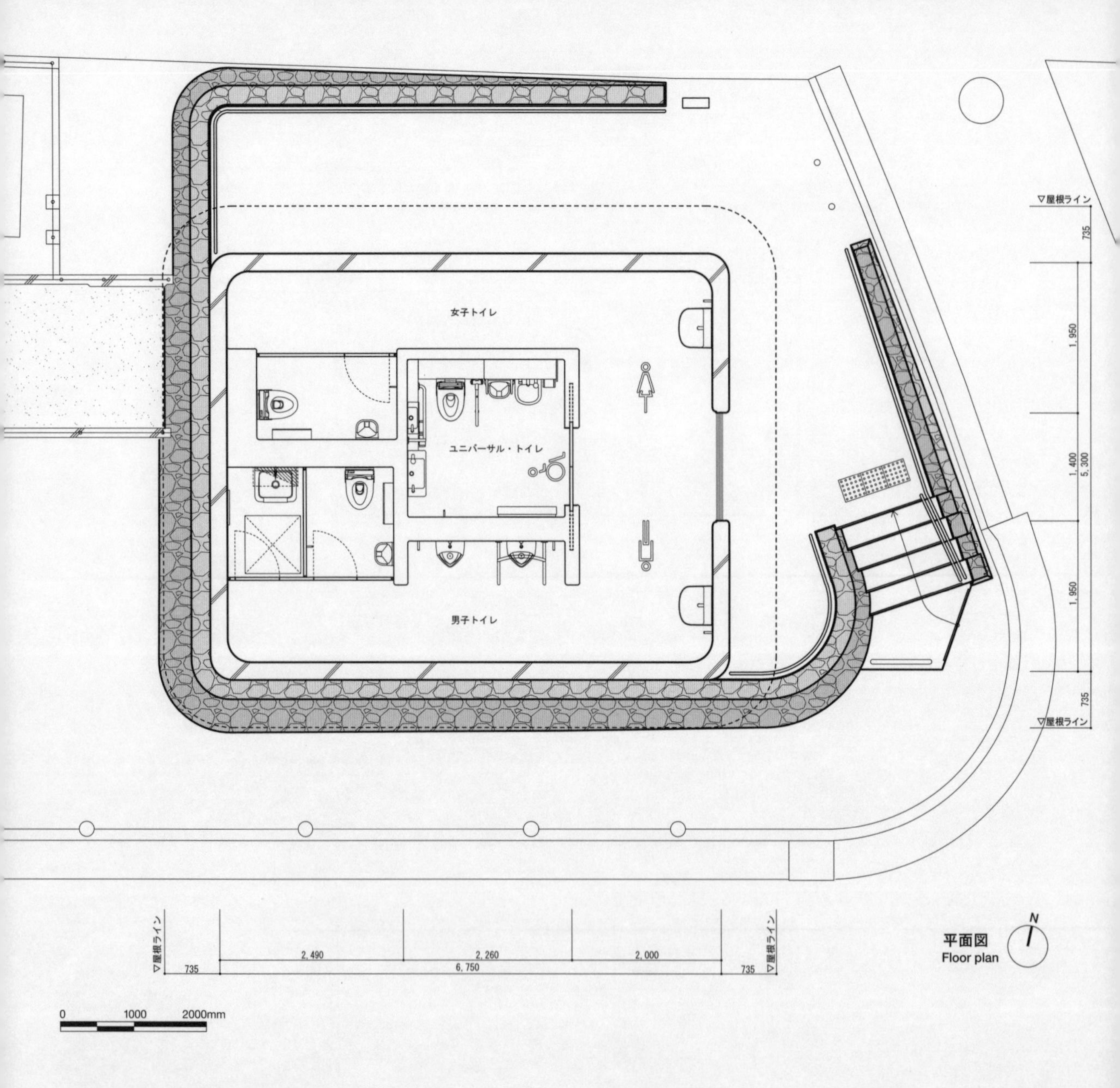

平面図
Floor plan

裏参道公衆トイレ P.142

URASANDO PUBLIC TOILET

【所在地】東京都渋谷区千駄ヶ谷4-28-1
【デザイン】Marc Newson Ltd／
マーク・ニューソン
【設計・施工】大和ハウス工業

【規模】敷地面積　95.97m²／
建築面積　35.62m²／延床面積　35.62m²
建蔽率　37.12%（許容: 80%）
【寸法】最高高　7,370mm／
軒高　4,560mm／
天井高　2,400mm／2,100mm
【敷地条件】商業地域　防火地域
60m高度地区
【構造】鉄筋コンクリート造　一部鉄骨造
【施工期間】2021年6月～2023年1月

【外部仕上げ】
屋根: 銅板 t=0.4mm 一文字葺
外壁: コンクリート化粧打ち放し仕上げ
外構: コンクリート刷毛引き仕上げ

【内部仕上げ】
床: ウレタン系弾性塗床材塗装
壁: 人工大理石／モルタル、耐水石膏ボード
ウレタン塗装
天井: ケイ酸カルシウム板　EP塗装

【主な使用機器】（TOTO）
壁掛大便器: UAXC1CS2A1／ウォシュレット
（エコリモコン）: TCF5840PN／壁掛小便器:
UFH500／スタッフ用手洗器:LS850DSA
コンパクト・バリアフリートイレパック:
UADBK61L1A1ASN2WA／ベビーシート:
YKA25S／ベビーチェア: YKA15S

Location
4-28-1 Sendagaya, Shibuya-ku, Tokyo
Design
Marc Newson Ltd./ Marc Newson
Design and construction
Daiwa House Industry

Size
Site area 95.97 m²/
Building area 35.62 m²/
Total floor area 35.62 m²/
Building coverage ratio 37.12%
(Maximum allowable ratio 80%)
Dimension
Maximum building height 7,370 mm/
Eave height 4,560 mm/
Ceiling height 2,400 mm/ 2,100 mm
Site condition
Commercial District, Fire Prevention District, 60 m- Height Control District
Structure
Reinforced concrete construction, partially steel construction
Construction period
June 2021 - January 2023

Exterior finishes
Roof: Copper plate t=0.4 mm, Ichimonji flat seam roofing
Exterior wall: Exposed fair-faced concrete finish
Exterior: Brush-finished concrete
Interior finishes
Floor: Elastic urethane floor coating
Wall: Artificial marble/ mortar, waterproof plasterboard with urethane coating
Ceiling: Calcium silicate board with emulsion paint finish

Sanitary fixtures used
(TOTO products)
Toilet: UAXC1CS2A1/
Washlet (with Eco Remote Controller): TCF5840PN/
Urinal: UFH500/ Hand wash basin: LS850DSA/
Wheelchair accessible toilet unit: UADBK61L1A1ASN2WA/
Baby changing station: YKA25S/
Baby seat: YKA15S

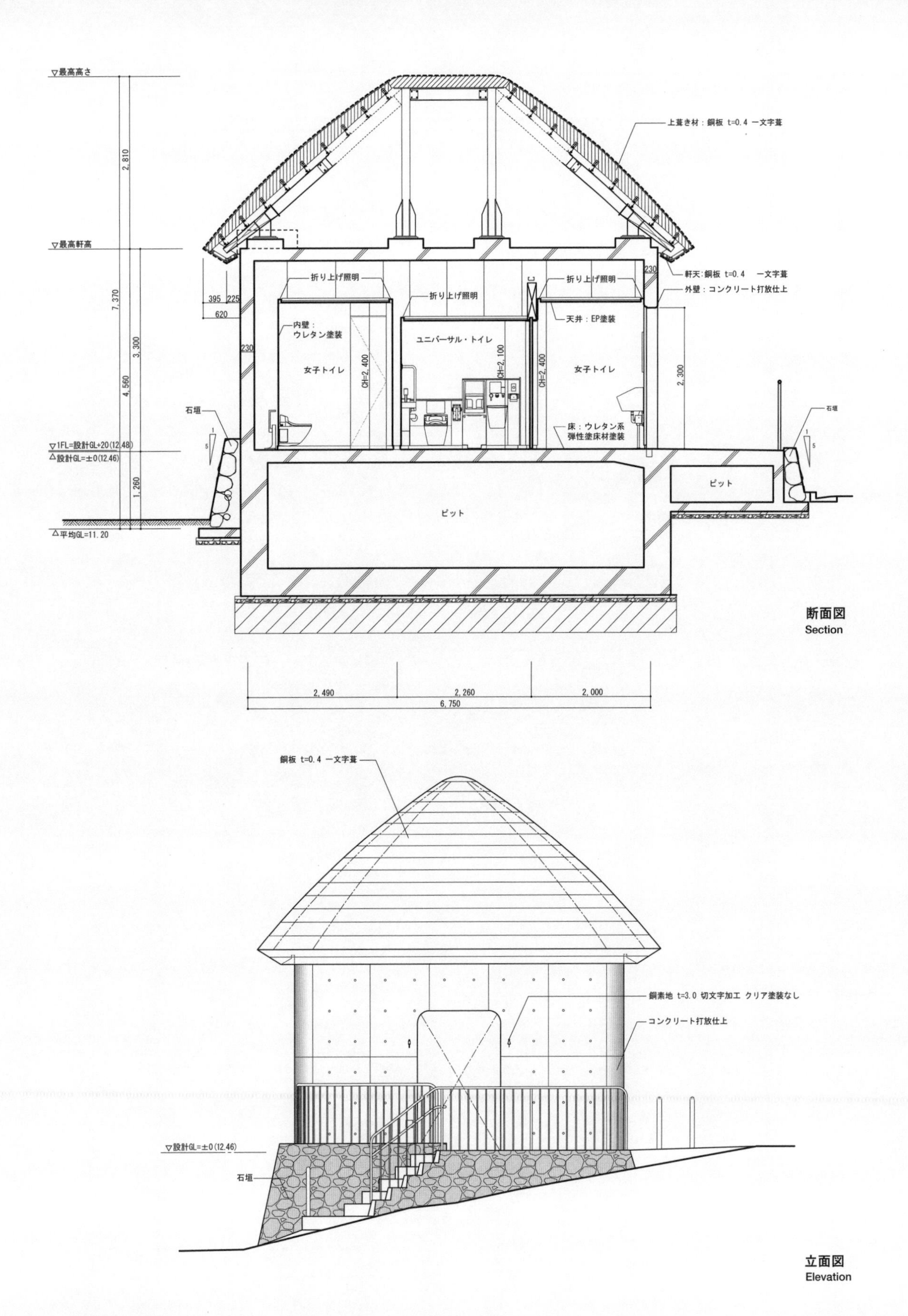

断面図
Section

立面図
Elevation

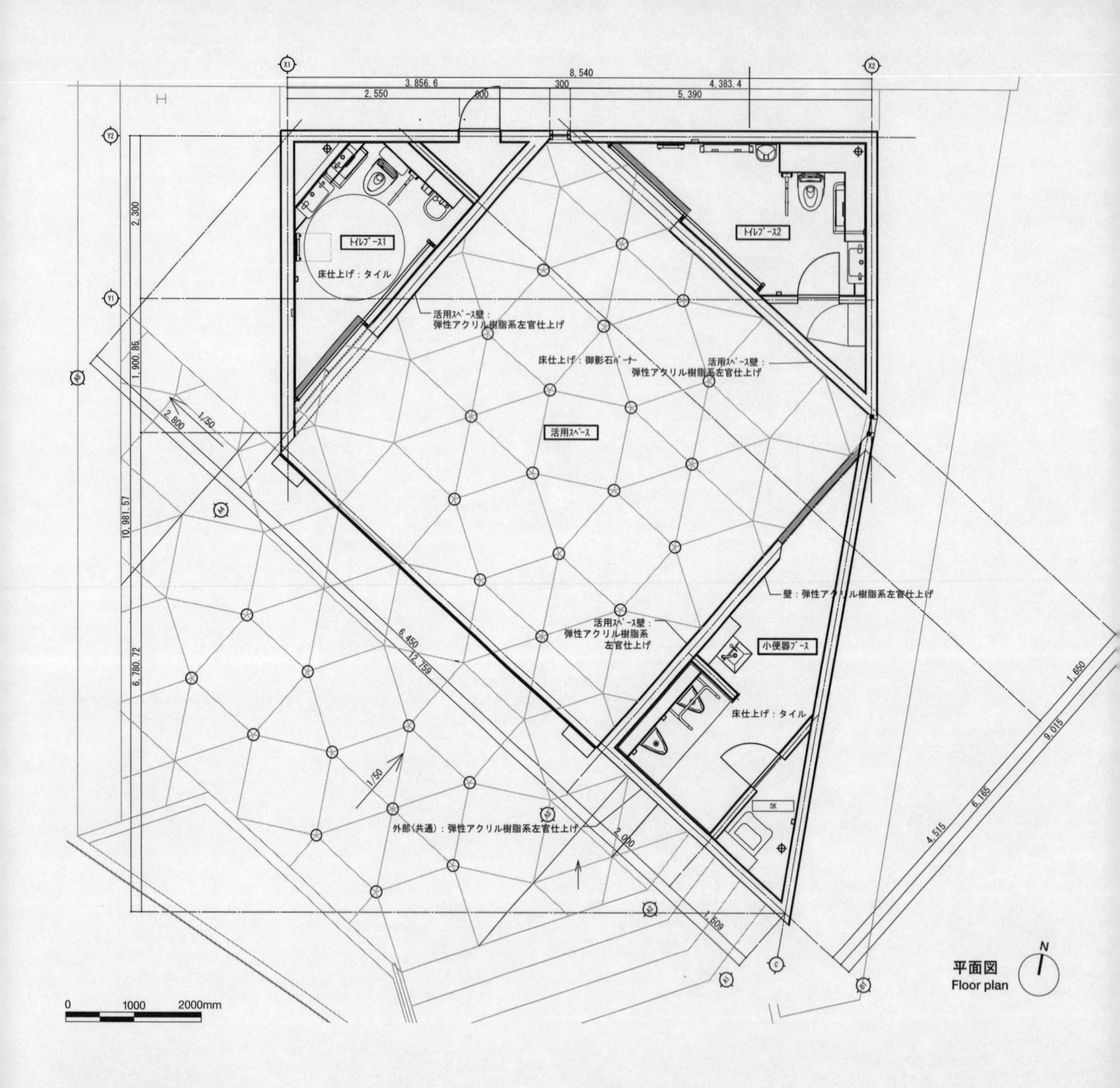

平面図
Floor plan

幡ヶ谷公衆トイレ P.152

HATAGAYA PUBLIC TOILET

【所在地】東京都渋谷区幡ヶ谷3-37-8
【デザイン】マイルス・ペニントン
今井公太郎　本間健太郎　山田将太郎
伊東優　国枝歓　クリスチャン・フェルスナー
磯部宏太　内倉悠
【構造デザイン協力】福島佳浩
【設計・施工】大和ハウス工業

【規模】敷地面積　170.02m²／
建築面積　65.60m²／延床面積　65.60m²
建蔽率　38.58%（許容: 80%）
【寸法】最高高　5,250mm／
軒高　5,150mm
【敷地条件】第二種住居地域・防火地域
40m高度地区
【構造】鉄筋コンクリート造
【施工期間】2021年10月～2023年2月

【外部仕上げ】
屋根: 超速硬化ウレタン防水
外壁: 弾性アクリル樹脂系多意匠装飾仕上塗材
開口部: スチールドア
外構: 御影石　ベンチ: 座面木製（イペ）
【内部仕上げ】
床: 御影石、タイル、防塵塗料　壁・天井: 弾性アクリル樹脂系多意匠装飾仕上塗材

【主な使用機器】（TOTO）
壁掛大便器: UAXC3CS1／パブリックコンパクト便器: CFS498BCK／ウォシュレット（エコリモコン）: TCF5840AUPN／壁掛小便器: XPU21A／壁掛洗面器: L710CM／台付自動水栓: TLE26506J／コンパクトオストメイトパック: UAS81RSB2NW／ベビーシート: YKA25S／ベビーチェア: YKA15S／フィッティングボード: YKA41

Location
3-37-8 Hatagaya, Shibuya-ku, Tokyo
Design
Miles Pennington, Kotaro Imai, Kentaro Honma, Shotaro Yamada, Yu Ito, Kan Kunieda, Christian Felsner, Kota Isobe, Yu Uchikura
Structural design cooperation
Yoshihiro Fukushima
Design and construction
Daiwa House Industry

Size　Site area 170.02 m²/
Building area 65.60 m²/
Total floor area 65.60 m²/
Building coverage ratio 38.58%
(Maximum allowable ratio: 80%)
Dimension
Maximum building height 5,250 mm/
Eave height 5,150 mm
Site condition
Category 2 Residential District,
Fire Prevention District,
40 m-Height Control District
Structure
Reinforced concrete construction
Construction period
October 2021- February 2023

Exterior finishes
Roof: Ultra-fast curing urethane waterproofing
Exterior wall: Elastic acrylic resin coating for multifunctional decorative finishes
Opening: Steel door
Exterior: Granite
Bench: Wooden seat (Ipe)
Interior finishes
Floor: Granite and tile with dustproof coating
Wall/ceiling: Elastic acrylic resin coating for multifunctional decorative finishes

Sanitary fixtures used
(TOTO products)
Toilet: UAXC3CS1/ Toilet: CFS498BCK/
Washlet (with Eco Remote Controller): TCF5840AUPN/ Urinal: XPU21A/
Wash basin: L710CM/
Automatic faucet: TLE26506J/
Ostomate sink: UAS81RSB2NW/
Baby changing station: YKA25S/
Baby seat: YKA15S/
Changing board: YKA41

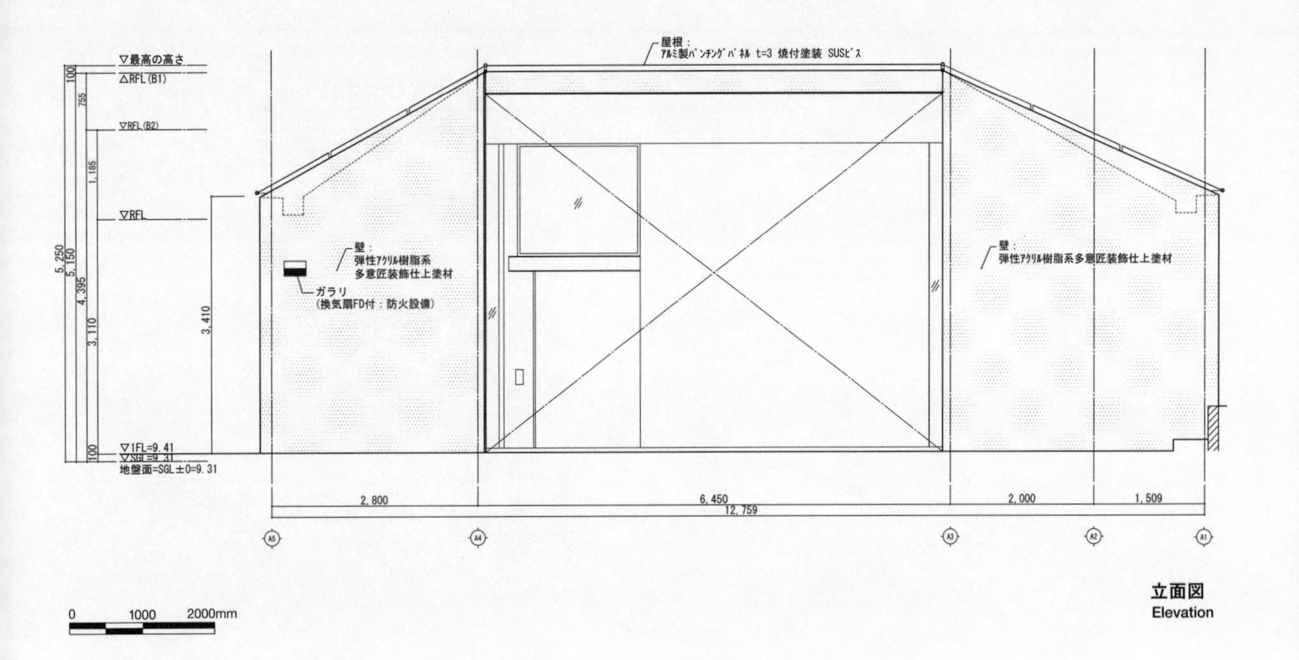

立面図
Elevation

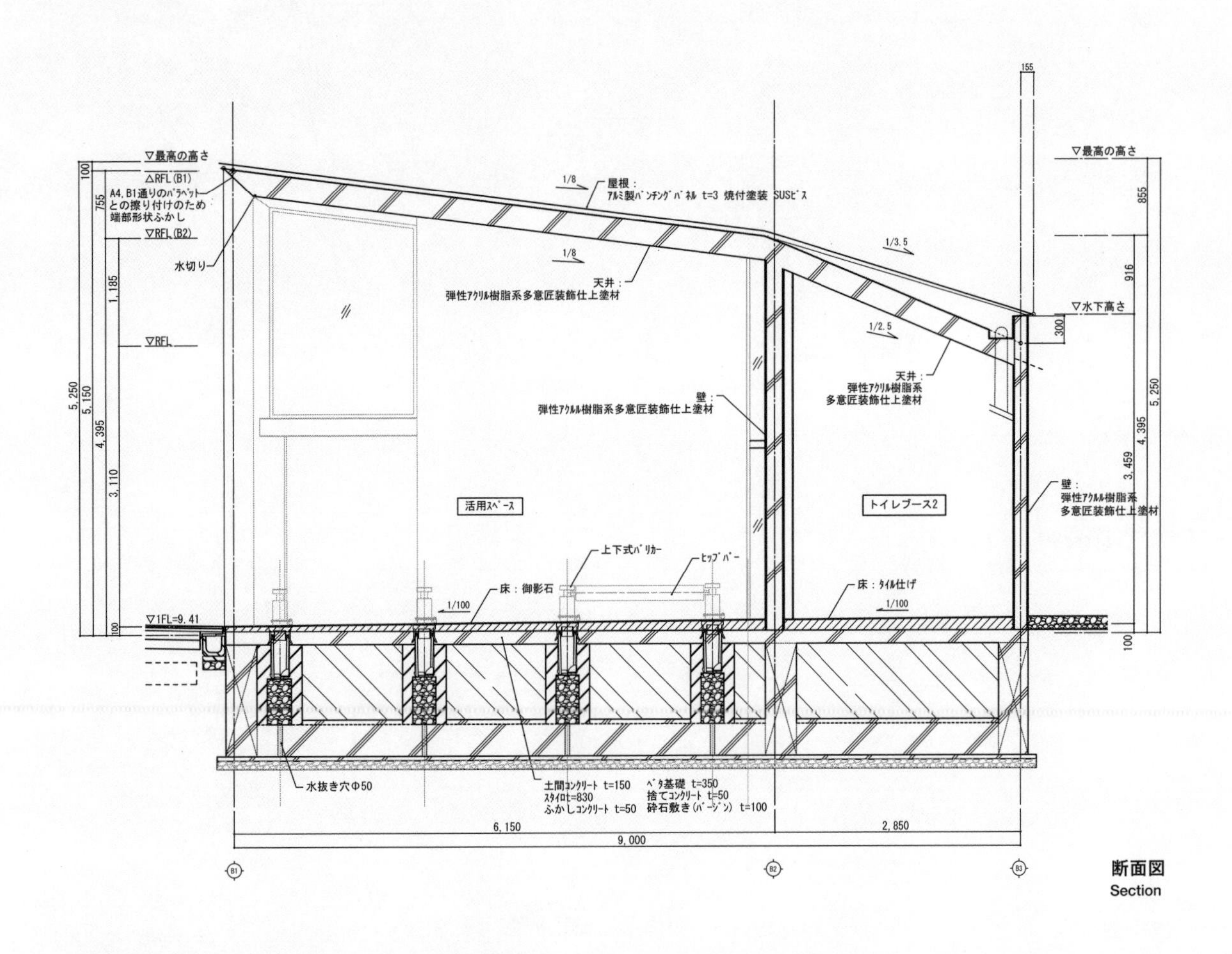

断面図
Section

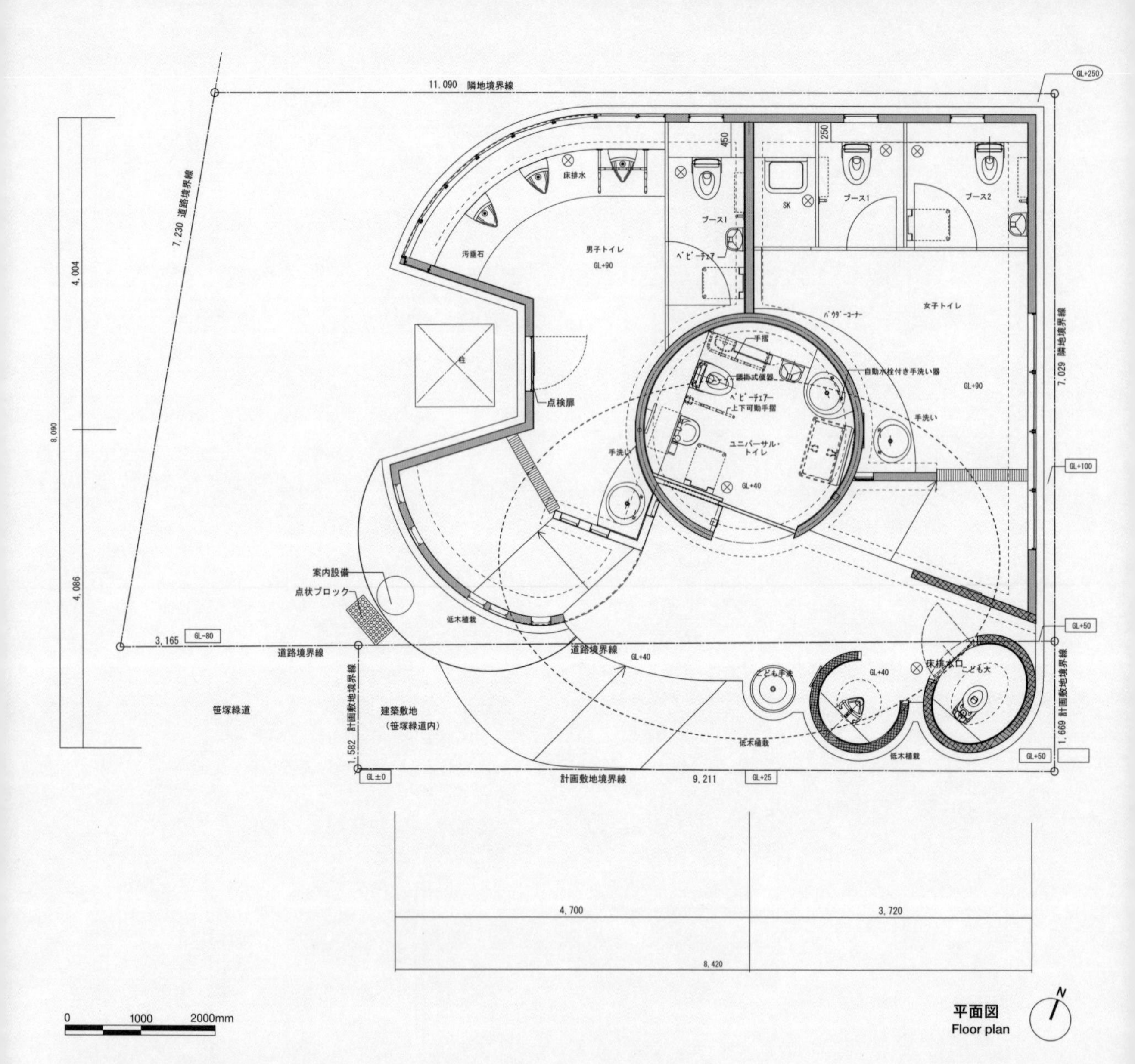

平面図
Floor plan

笹塚緑道公衆トイレ P.162

SASAZUKA GREENWAY PUBLIC TOILET

【所在地】東京都渋谷区笹塚1-29
【デザイン】設計事務所ゴンドラ／小林純子
【構造監修】梅沢建築構造研究所
【設計・施工】大和ハウス工業

【規模】敷地面積　97.89m²／建築面積　51.17m²／延床面積　51.17m²
建蔽率　52.27%（許容: 80%）
【寸法】最高高　4,050mm／軒高　3,450mm／階高　3,450mm／天井高　3,300mm
【敷地条件】商業地域　防火地域　50m高度地区
【構造】鉄骨造（サンドイッチパネル工法）
【施工期間】2021年1月〜2023年3月

【外部仕上げ】
屋根: 耐候性鋼板　ウレタン塗装
外壁・開口部・建具: 耐候性鋼板（一部サッシ　ステンレス焼付仕上）
外構: 砂利　洗い出し仕上

【内部仕上げ】
床: 磁器質タイル貼
壁: 耐候性鋼板　一部モザイクタイル貼
天井: ケイカルt=6mmの上EP塗装

【主な使用機器】（TOTO）
壁掛大便器: UAXC3CS1／ウォシュレット（エコリモコン）: TCF5840AUP*系／
自動洗浄小便器: UFS900JCS／
ハイドロセラ・フロアPU: AB690系／
台付自動水栓: TLE26SS1A／
コンパクトオストメイトパック: UAS81RSB2NW
幼児用大便器: CS300B／
幼児用小便器: U310GY／
ベビーシート: YKA25S／
ベビーチェア: YKA15S

Location
1-29 Sasazuka, Shibuya-ku, Tokyo
Design
GONDOLA/ Junko Kobayashi
Structural engineering supervision
Umezawa Structural Engineers

Design and construction
Daiwa House Industry

Size Site area 97.89 m²/ Building area 51.17 m²/ Total floor area 51.17 m²/ Building coverage ratio 52.27% (Maximum allowable ratio: 80%)
Dimension
Maximum building height 4,050 mm/ Eave height 3,450 mm/ Floor height 3,450 mm/ Ceiling height 3,300 mm
Site condition
Commercial District, Fire Prevention District, 50 m- Height Control District
Structure
Steel construction (sandwich panel method)
Construction period
January 2021 - March 2023

Exterior finishes
Roof: Weather-resistant steel plate with urethane coating
Exterior wall/ opening/ fittings: Weather-resistant steel plate, partially with stainless steel window frames with baked finish
Exterior: Pebble and exposed aggregate finish
Interior finishes
Floor: Porcelain tile
Wall: Weather-resistant steel plate, partially mosaic tile
Ceiling: Calcium silicate board t=6 with emulsion paint finish

Sanitary fixtures used
(TOTO products)
Toilet: UAXC3CS1/ Washlet (with Eco Remote Controller): TCF5840AUP*/ Urinal: UFS900JCS/
Large ceramic tile (for use under urinals): AB690/ Automatic faucet: TLE26SS1A/ Ostomate sink: UAS81RSB2NW/ Infant toilet: CS300B/ Infant urinal: U310GY/ Baby changing station: YKA25S/ Baby seat: YKA15S

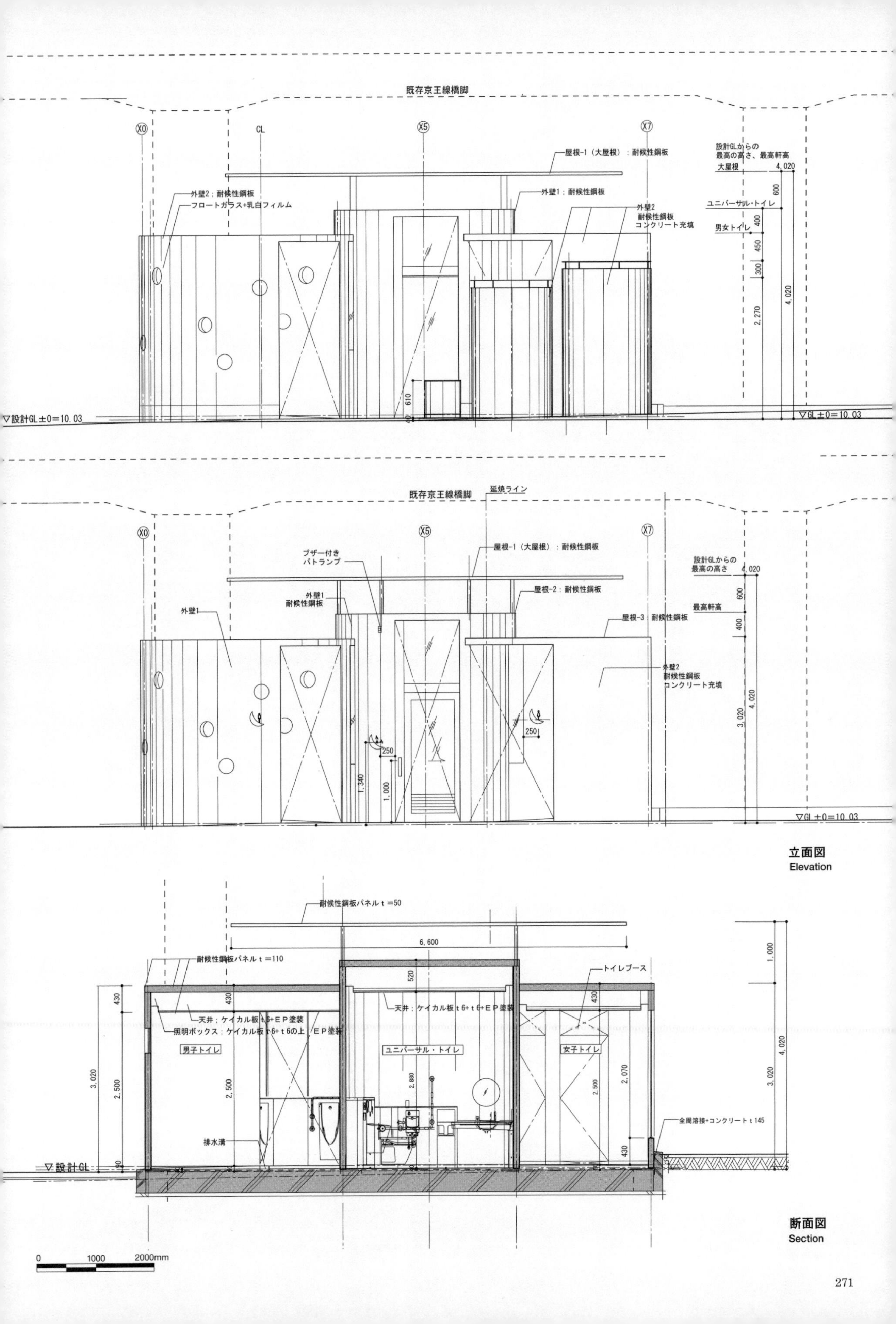

既存京王線橋脚
X0
CL
X5
X7
屋根-1（大屋根）：耐候性鋼板
外壁2：耐候性鋼板
フロートガラス+乳白フィルム
外壁1：耐候性鋼板
外壁2
耐候性鋼板
コンクリート充填
設計GLからの
最高の高さ、最高軒高
大屋根
4,020
ユニバーサル・トイレ
男女トイレ
600
400
450
300
2,270
4,020
610
40
▽設計GL±0=10.03
▽GL±0=10.03
既存京王線橋脚
延焼ライン
X0
X5
X7
屋根-1（大屋根）：耐候性鋼板
ブザー付き
パトランプ
外壁1
耐候性鋼板
屋根-2：耐候性鋼板
外壁1
屋根-3 耐候性鋼板
設計GLからの
最高の高さ
4,020
最高軒高
600
400
外壁2
耐候性鋼板
コンクリート充填
250
250
1,340
1,000
3,020
4,020
▽GL±0=10.03
立面図
Elevation
耐候性鋼板パネル t=50
6,600
耐候性鋼板パネル t=110
トイレブース
430
430
520
430
天井：ケイカル板 t6+t6+EP塗装
照明ボックス：ケイカル板 t6+t6の上 EP塗装
天井：ケイカル板 t6+t6+EP塗装
男子トイレ
ユニバーサル・トイレ
女子トイレ
3,020
2,500
2,500
2,880
2,500
2,070
1,000
3,020
4,020
全周溶接+コンクリート t145
排水溝
430
90
▽設計GL
断面図
Section
0
1000
2000mm

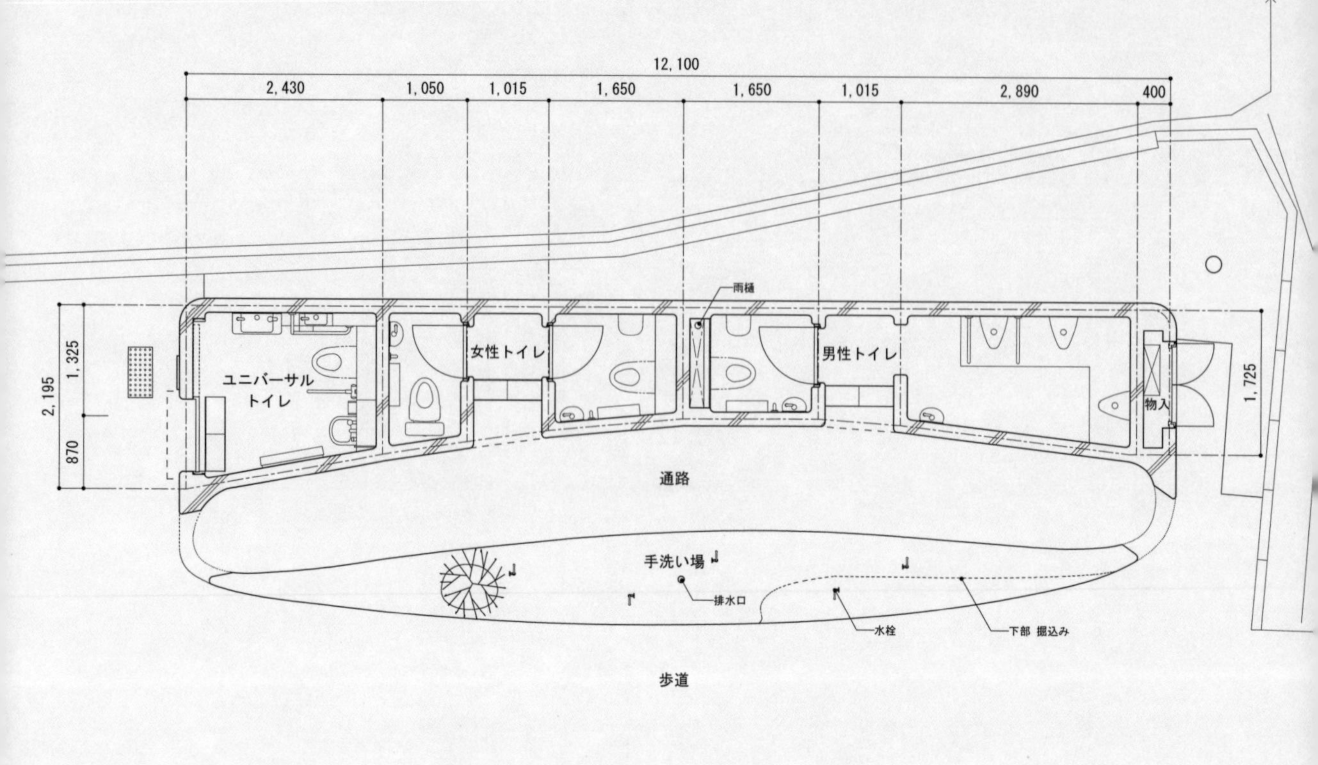

平面図
Floor plan

西参道公衆トイレ P.172
NISHISANDO PUBLIC TOILET

【所在地】東京都渋谷区代々木3-27-1
【デザイン】藤本壮介建築設計事務所／
藤本壮介　保坂整　國清尚之　窪田啓吾
本間新太郎
【設計・施工】大和ハウス工業

【規模】建築面積　19.24m²／
延床面積　19.24m²
【寸法】最高高　2,950mm／
軒高　2,870mm／
天井高　2,100～2,250mm
【敷地条件】商業地域　防火地域
【構造】鉄筋コンクリート造
【施工期間】2021年6月～2023年3月

【外部仕上げ】
屋根・外壁: 超速硬化ウレタン防水の上、
光触媒コーティング
開口部: ステンレス自動ドア
スチールドア
外構: ウレタン樹脂系塗床
【内部仕上げ】
床: ウレタン樹脂系塗床
壁・天井: EP

【主な使用機器】(TOTO)
パブリックコンパクト便器: CFS498BLC／
壁掛大便器: UAXC3CS1／
ウォシュレット(エコリモコン): TCF5840AUPN
自動洗浄小便器: UFS900JS／
壁掛手洗器: LSE90BBSZ／
コンパクト・バリアフリートイレパック:
UADBK21L1A1ASN2WA／
ベビーシート: YKA25S／
ベビーチェア: YKA15S

Location
3-27-1 Yoyogi, Shibuya-ku, Tokyo
Design
Sou Fujimoto Architects/ Sou Fujimoto,
Sei Hosaka, Naoyuki Kunikiyo,
Keigo Kubota, Shintaro Honma
Design and construction
Daiwa House Industry

Size
Building area 19.24 m²/
Total floor area 19.24 m²
Dimension
Maximum building height 2,950 mm/
Eave height 2,870 mm/
Ceiling height 2,100 - 2,250 mm
Site condition
Commercial District,
Fire Prevention District
Structure
Reinforced concrete construction
Construction period
June 2021 - March 2023

Exterior finishes
Roof/ exterior wall: Ultra-fast
curing urethane waterproofing with
photocatalytic coating
Opening: Stainless steel automatic door
and steel door
Exterior: Urethane resin floor coating
Interior finishes
Floor: Urethane resin floor coating
Wall/ceiling: Emulsion paint finish

Sanitary fixtures used
(TOTO products)
Toilet: CFS498BLC/
Toilet: UAXC3CS1/
Washlet (with Eco Remote Controller):
TCF5840AUPN/
Urinal: UFS900JS/ Hand wash basin:
LSE90BBSZ/
Wheelchair accessible toilet unit:
UADBK21L1A1ASN2WA/
Baby changing station: YKA25S/
Baby seat: YKA15S

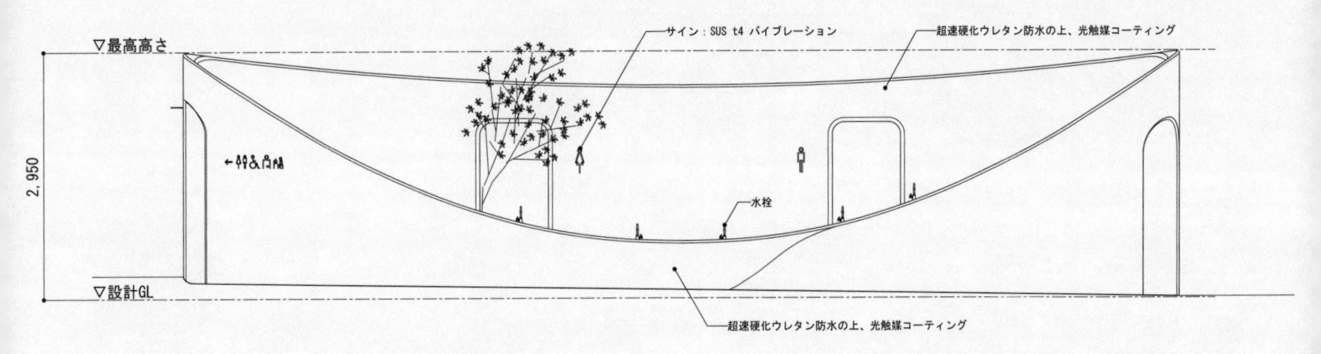

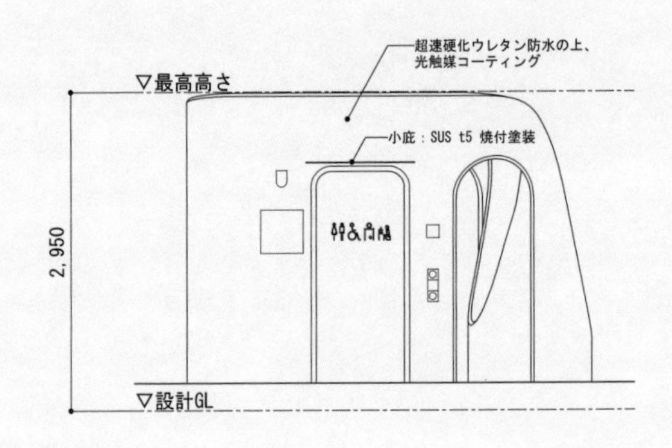

立面図
Elevation

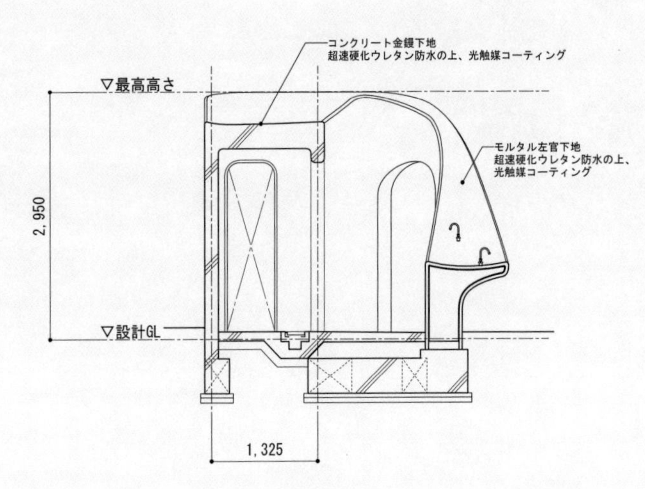

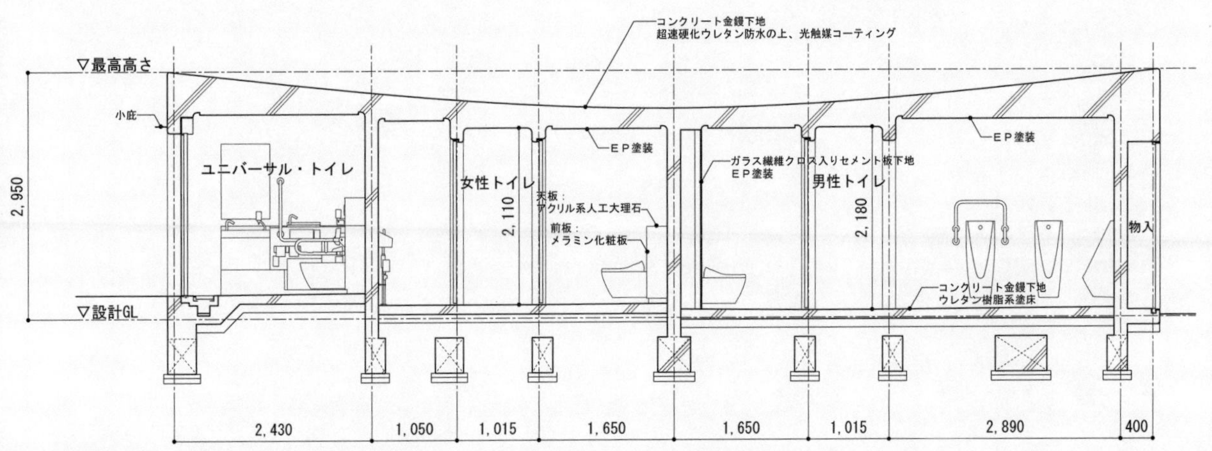

断面図
Section

0 1000 2000mm

LOCATION MAP

01 笹塚緑道公衆トイレ
SASAZUKA GREENWAY PUBLIC TOILET
渋谷区笹塚1丁目29番
1-29 Sasazuka Shibuya-ku
P.162

02 幡ヶ谷公衆トイレ
HATAGAYA PUBLIC TOILET
渋谷区幡ヶ谷3丁目37番8号
3-37-8 Hatagaya Shibuya-ku
P.152

03 七号通り公園トイレ
NANAGO DORI PARK PUBLIC TOILET
渋谷区幡ヶ谷2丁目53番5号
2-53-5 Hatagaya Shibuya-ku
P.122

04 西原一丁目公園トイレ
NISHIHARA ITCHOME PARK PUBLIC TOILET
渋谷区西原1丁目29番1号
1-29-1 Nishihara Shibuya-ku
P.62

05 西参道公衆トイレ
NISHISANDO PUBLIC TOILET
渋谷区代々木3丁目27番1号
3-27-1 Yoyogi Shibuya-ku
P.172

06 代々木八幡公衆トイレ
YOYOGI-HACHIMAN PUBLIC TOILET
渋谷区代々木5丁目1番2号
5-1-2 Yoyogi Shibuya-ku
P.112

07 はるのおがわコミュニティパークトイレ
HARU-NO-OGAWA COMMUNITY PARK PUBLIC TOILET
渋谷区代々木5丁目68番1号
5-68-1 Yoyogi Shibuya-ku
P.22

08 代々木深町小公園トイレ
YOYOGI FUKAMACHI MINI PARK PUBLIC TOILET
渋谷区富ヶ谷1丁目54番1号
1-54-1 Tomigaya Shibuya-ku
P.18

09 裏参道公衆トイレ
URASANDO PUBLIC TOILET
渋谷区千駄ヶ谷4丁目28番1号
4-28-1 Sendagaya Shibuya-ku
P.142

10
P.82
神宮前公衆トイレ
JINGUMAE PUBLIC TOILET
渋谷区神宮前1丁目3番14号
1-3-14 Jingumae Shibuya-ku
11
P.72
神宮通公園トイレ
JINGU-DORI PARK PUBLIC TOILET
渋谷区神宮前6丁目22番8号
6-22-8 Jingumae Shibuya-ku
12
P.92
鍋島松濤公園トイレ
NABESHIMA SHOTO PARK PUBLIC TOILET
渋谷区松濤2丁目10番7号
2-10-7 Shoto Shibuya-ku
13
P.42
東三丁目公衆トイレ
HIGASHI SANCHOME PUBLIC TOILET
渋谷区東3丁目27番1号
3-27-1 Higashi Shibuya-ku
14
P.32
恵比寿公園トイレ
EBISU PARK PUBLIC TOILET
渋谷区恵比寿西1丁目19番1号
1-19-1 Ebisu-Nishi Shibuya-ku
15
P.102
恵比寿駅西口公衆トイレ
EBISU STATION, WEST EXIT PUBLIC TOILET
渋谷区恵比寿南1丁目5番8号
1-5-8 Ebisu-Minami Shibuya-ku
16
P.52
恵比寿東公園トイレ
EBISU EAST PARK PUBLIC TOILET
渋谷区恵比寿1丁目2番16号
1-2-16 Ebisu Shibuya-ku
17
P.132
広尾東公園トイレ
HIROO HIGASHI PARK PUBLIC TOILET
渋谷区広尾4丁目2番27号
4-2-27 Hiroo Shibuya-ku
新宿区
Shinjuku City
千駄ヶ谷駅
Sendagaya Sta.
代々木駅
Yoyogi Sta.
北参道駅
Kita-Sando Sta.
東京メトロ副都心線
Tokyo Metro Fukutoshin Line
明治神宮
Meiji Jingu
谷
区
hibuya City
原宿駅
Harajuku Sta.
代々木公園
Yoyogi Park
東京メトロ千代田線
Tokyo Metro Chiyoda Line
明治神宮前駅
Meiji-jingumae Sta.
港区
Minato City
JR山手線
Yamanote Line
東京メトロ銀座線
Tokyo Metro Ginza Line
六本木通り
Roppongi-dori St.
渋谷駅
Shibuya Sta.
明治通り
Meiji-dori Ave.
神泉駅
Shinsen Sta.
京王井の頭線
Keio Inokashira Line
広尾駅
Hiroo Sta.
東京メトロ日比谷線
Tokyo Metro Hibiya Line
東急東横線
Tokyu Toyoko Line
恵比寿駅
Ebisu Sta.
目黒区
Meguro City
代官山駅
Daikan-yama Sta.

CREDITS

翻訳	**Translator**
英訳　坂本和子	English Translation: Kazuko Sakamoto
英文校正　織部晴崇 （坂本和子による英訳部分）	English Proofreading: Harutaka Oribe （All English texts translated by Kazuko Sakamoto.）
和訳　柴田元幸	Japanese Translation: Motoyuki Shibata
校正	**Proofreading**
株式会社鷗来堂	Ouraidou K.K.
協力	**Cooperation**
日本財団	The Nippon Foundation
大和ハウス工業株式会社	Daiwa House Industry Co., Ltd.

The Tokyo Toilet

First edition published in Japan on October 20, 2023
First edition, third printing published on November 15, 2024

Author: Tami Okano [Text] Satoshi Nagare [Photo]
Publisher: Akira Watai
TOTO Publishing [TOTO LTD.]
TOTO Nogizaka Bldg. 2F, 1-24-3 Minami-Aoyama
Minato-ku, Tokyo 107-0062, Japan
[Sales] Telephone: +81-3-3402-7138
Facsimile: +81-3-3402-7187
[Editorial] Telephone: +81-3-3497-1010
URL: https://jp.toto.com/publishing
Book Designer: Takashi Shimada
Editor: Masashi Matsuie, Yuri Kitamoto [Tsuru & Hana Co.]
Printer: TOPPAN Colorer Inc.

ISBN978-4-88706-404-1

The Tokyo Toilet

2023年10月20日　初版第1刷発行
2024年11月15日　初版第3刷発行

著者　岡野 民［文］ 永禮 賢［写真］
発行者　渡井 朗
発行所　TOTO出版［TOTO株式会社］
〒107-0062 東京都港区南青山1-24-3
TOTO乃木坂ビル2F
［営業］TEL: 03-3402-7138 FAX: 03-3402-7187
［編集］TEL: 03-3497-1010
URL: https://jp.toto.com/publishing
デザイン　島田 隆
編集　松家仁之　北本侑理［株式会社つるとはな］
印刷・製本　TOPPANクロレ株式会社

ISBN978-4-88706-404-1